44 Advances in Biochemical Engineering
Biotechnology

Managing Editor: A. Fiechter

Bioreactor
Systems and Effects

With contributions by
R. F. Bliem, H. N. Chang, H. W. Doelle,
S. Furusaki, P. F. Greenfield, M. R. Johns,
H. W. D. Katinger, K. Konopitzky, V. Křen,
J. C. Merchuk

With 76 Figures and 30 Tables

Springer-Verlag Berlin
Heidelberg GmbH

ISBN 978-3-662-15021-4   ISBN 978-3-540-47400-5 (eBook)
DOI 10.1007/978-3-540-47400-5

Originally published by Springer-Verlag Berlin Heidelberg New York in 1991.

Softcover reprint of the hardcover 1st edition 1991

Typesetting: Th. Müntzer, Bad Langensalza; Printing: Heenemann, Berlin;

52/3020-543210 — Printed on acid-free paper

Managing Editor

Professor Dr. A. Fiechter
Institut für Biotechnologie, Eidgenössische Technische Hochschule
ETH — Hönggerberg, CH-8093 Zürich

Editorial Board

Table of Contents

Industrial Animal Cell Reactor Systems: Aspects of Selection and Evaluation

R. Bliem
Bristol-Myers Squibb Co., Syracuse, NY USA
K. Konopitzky
Bender GesmbH, Vienna, Austria
H. Katinger
IAM, University of Agriculture, Vienna, Austria

Advances in Biochemical Engineering
Biotechnology, Vol. 44
Managing Editor: A. Fiechter
© Springer-Verlag Berlin Heidelberg 1991

1 Introduction

The selection, evaluation and development of animal cell reactor systems for industrial applications are, under ideal conditions, aimed at determining an economic and quality optimum for the interactive factors for biology, chemistry, engineering and operations. In this paper we will present an overview of some of the considerations pertaining to these factors and discuss some of the topical aspects of reactor design based on the current state of technology.

The prominence of animal cell technology is largely based on the fact that it is still the only vehicle enabling the correct translation of genes into the complex structures required for functional fidelity (this is not to say that structural alterations necessarily entail functional changes).

Cultured cells vary in their response to the combined effects of physical and chemical variables, so that the optimization of a medium, for example, should therefore also take into account the reactor configuration and operating conditions, and vice versa.

Similarly, cellular activities determine the consumption and secretion of substances which in turn will affect the cellular activities, thus forming a mutually dependent feed-back loop.

2 Complexity of Cells, Reactors and Operations

The trend towards using conventional cell lines, such as Vero, BHK and CHO [1–3] reflects an uncertainty with respect to this technology. Their widespread use cannot be explained on the basis of any inherent benefits of these lines over other potentially available lines. A driving force for their selection as production lines is the availability and participation in knowledge presented in the public domain. The choice of such common lines may also reduce the potential risk and effort involved in the course of product registration. This "bandwagon effect" is similar to that observed with E. coli, which was stylized into the standard organism for recombinant DNA work. Ignoring the disadvantages inherent in some of these conventional lines, they do offer the very practical, but short-term benefit of a quick start into a new technology development, by following procedures already established in the literature.

On the other hand, the strategy of employing the same cell line(s) (not necessarily the conventional lines) for a variety of products has a strong economic appeal; it would not only simplify the effort required in the development of the different production processes, but also reduce the work load expended in quality control and risk assessment (i.e. regulatory requirements).

The widespread application of a few "conventional" cell lines as fusion partners for lymphocytes (e.g. SP2/0, NS-1, [4]) has led to hybridomas with similar culture characteristics. To a certain degree this is enabling the development of processes with similar technological features and requirements. However, these technical commonalities are not sufficient to support direct comparisons of the performance characteristics of different reactor systems using different hybridoma lines. The physiological, cultural and product formation characteristics still vary significantly between cell lines.

These qualitative and quantitative differences between cell lines used in the investigations of reactor systems hampers the direct comparison of reactor performance.

The process characteristics described below for individual reactors and conditions must be interpreted critically in the context of what has been discussed so far. The generalizations are intended to serve merely as a guide through the maize of reactors and operating conditions; it is by no means comprehensive nor definitive.

3 Reactor Systems, Elements and Operating Features

In principle suspension-type cells are generally propagated as single cell or aggregate suspensions, whereas anchorage-type cells require a solid attachment matrix to support propagation. This matrix may be a static surface, such as a Roux-type flask, or may also be held in liquid suspension, as is the case with microcarriers and macroporous carriers (see below). In addition to providing an environment "low" in hydrodynamic shear stress, we must satisfy the mass transfer requirements for oxygen, nutrients and products, in order to propagate and maintain the culture.

These basic requirements are then improved upon by the introduction of process control features. Both mass transfer and process control are easier to implement in reactors in which the cells are present in homogeneous suspension.

This was first achieved, on an industrial scale, with standard (microbial) Stirred Tank Reactors (STRs) (Table 3) for the production of Foot and Mouth Disease vaccine [5]. The experience gained from their extensive application in the bioindustry has compensated for complex scale-up characteristics. To-date, the largest reactors of this kind used for animal cells have working volumes of between 5000 and 10000 l.

Prompted by the simplified scale-up and operating features a number of companies have instead adopted for the Airlift Reactor configuration (ALR) (Fig. 1 and Table 1), which has now been used for more than a decade for the industrial propagation of hybridomas and other suspension cells [6, 7].

Instead of using an impeller the cells are held in suspension by sparging alone, which also serves to oxygenate the medium. The oxygenation efficiency has been shown to

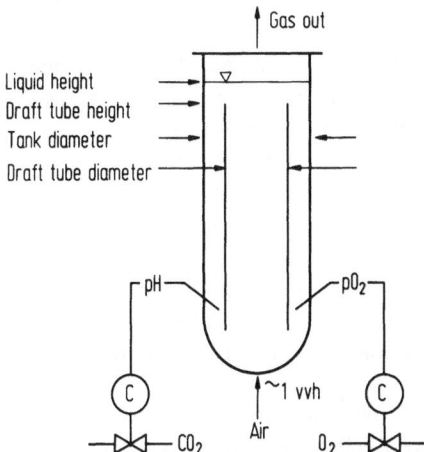

Fig. 1. Basic design of the airlift reactor configuration. Scale-up based on:

$Hl/Dt > 6$

$(Dt^2 - Dd^2)/Dd^2 \sim 1$

Hl ... Liquid Height Dt ... Tank Diameter

Hd ... Draft Tube Dd ... Draft Tube

      Height   Diameter

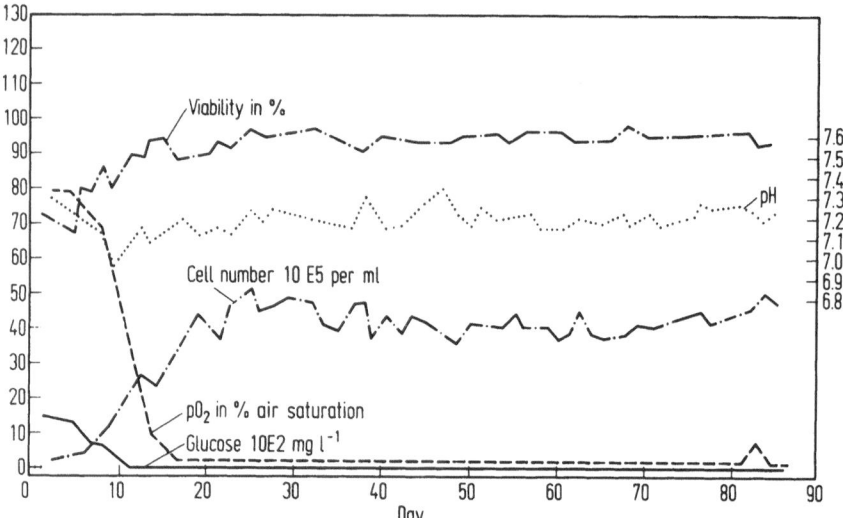

Fig. 2. Continuous propagation of Namalva cells in an 80 l airlift reactor

improve on scale-up and, when using air, is generally sufficient to support cell densities of up to 4 E6 cells per ml; increasing the oxygen concentration in the sparge gas potentially enables increases of up to E7 cells per ml, which of course also depends on the properties of the cell line [8]. Figure 2 presents data from a continuous culture of Namalva cells (used in the production of Alpha Interferon) in an 80 l airlift reactor.

Disadvantages of Airlift Reactors arise from the possible cell damage associated with direct sparging, particularly with microcarrier-based cultures. Operations may also be complicated by the coupling of the agitation and oxygenation functions, and by the probable need for the addition of an anti-foam agent (although this is not always necessary).

Bubble-associated shear damage is avoided in Fluidized Bed Reactors as the cells are propagated within macroporous carrier particles, which are held in liquid suspension. In addition, these reactors may be oxygenated bubble-free using gas permeable membranes.

The concept of the Fluidized Bed Reactor (FBR; Fig. 3) has been applied to many areas of the chemical and microbial processing industry (see [9, 10] and Table 2). In principle the fluidized bed reactor consists of a cylindrical or conical reactor vessel; medium is pumped through the vessel in an upward flow, thereby suspending solid carrier particles in which most (but not all) cells are entrapped. Particle fluidization is determined by such parameters as the particle (material) density relative to the liquid, by the size, shape and the uniformity of the particle, and the liquid velocity. Sparging will also significantly affect the performance characteristics.

The liquid velocity, which should not be less than 1 cm s^{-1}, determines the medium gradient across the length of the reactor and the mass transfer efficiency through the particle.

Solid carrier matrices have been developed so that anchorage dependent cells may be grown in homogeneous suspension. However, cells grown on the surface of such

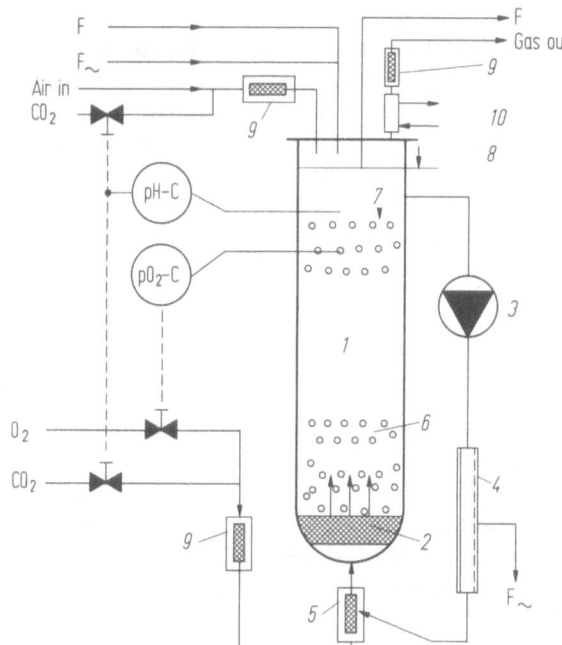

Fig. 3 Configuration features of a fluidized bed reactor.

1 Fluidized bed reactor
2 Flow-distributor
3 Low shear pump
4 Cross flow ultrafiltration module
5 Gas microsparger
6 Porous carriers
7 Level of fluidized bed
8 Liquid level
9 Air filter
10 Reflux cooler

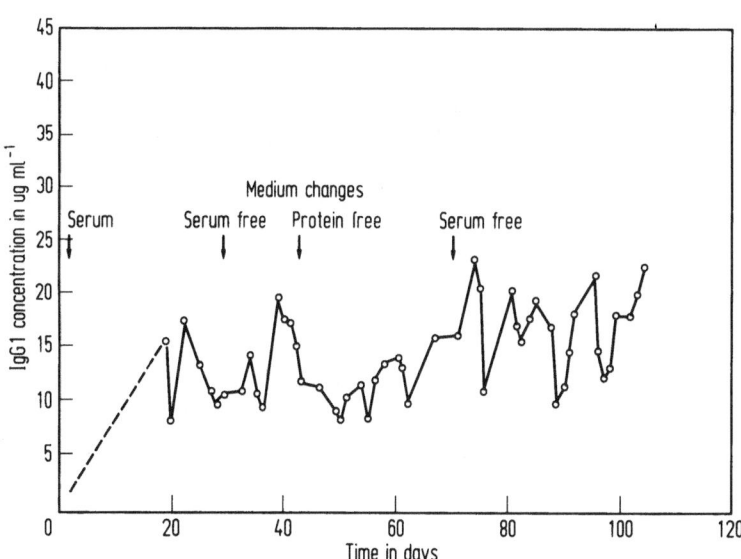

Fig. 4. Continuous propagation of murine hybridoma cell line in a scaled-down glass bead packed bed reactor.
Total bed volume: 1 liter; ~ 35% void volume between beads; medium feed rate: ~ 18 volumes per void volume per d; medium recirculation rate: 1.5 to 2 volumes per void volume per min

matrices (microcarriers) are generally sensitive to bubble associated damage, whereas cells grown within the matrices of macroporous carriers are protected from these effects (macroporous carriers are generally larger than microcarriers, $> 500\ \mu$ vs. $< 200\ \mu$ diameter respectively). On the other hand the cells may encounter mass transfer limitations, although experimental data on diffusion characteristics within such macrocarriers is only beginning to emerge. These carriers must of course consist of an inert material such as glass (Katinger, unpublished), alginate [11], collagen (Verax Corp., USA; Perstorp Biolytica AB, Sweden). With collagen carriers, care must be taken to ensure that the collagen is neither enzymatically degraded nor introduces contaminants into the raw product (see [12] for immobilization techniques).

The volume fraction of the suspended carriers can comprise up to 90% of the working volume of the reactor, thereby enabling cell concentrations which are much higher than those typical for stirred tank reactors.

With appropriate carriers the fluidized bed system can be scaled up as a unit process to any currently relevant capacity using standard bioprocess equipment. Although there is still room for improvement, especially with respect to the carriers, the fluidized bed concept comprises many of the features desirable for animal cell propagation, such as high cell retention and perfusion capabilities, simple scale-up parameters and bubble-free operation. The mixed particle suspension also facilitates process control and uniform medium dispersion.

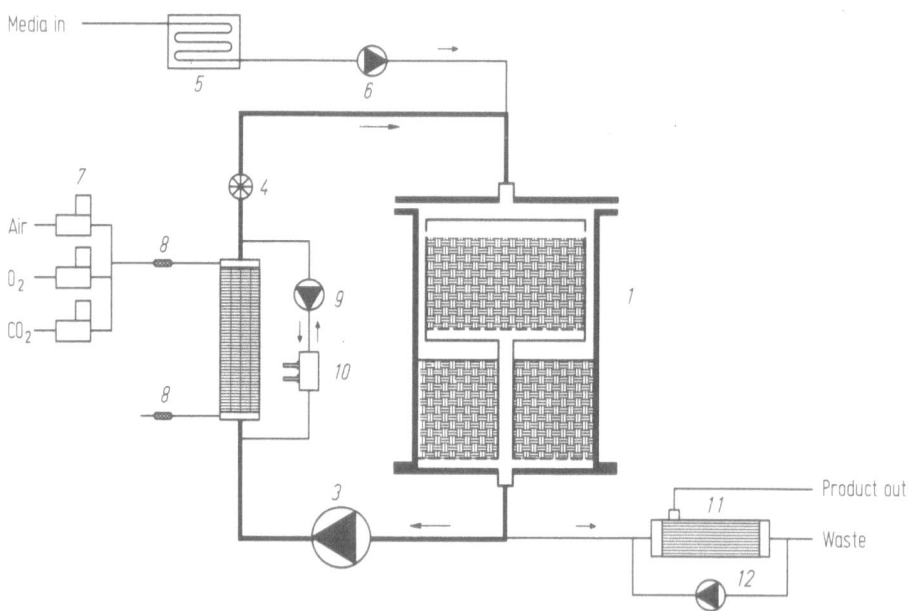

Fig. 5. Packed bed reactor with improved scale-up features

1 Packed bed reactor	7 Gas mass flow meter-controllers
2 Silicone membrane lung	8 Gas filters (0.2 μ)
3 Low shear pump	9 Reversable sample pump
4 Mass flow meter	10 pO₂ & pH sensors
5 Media pre-heater	11 Cross flow clarification module
6 Media pump	12 Recirculation pump

Typical problem areas concerning this reactor type, apart from the carriers, are the design and operation of the recirculation pump and of the oxygenator (fouling problems, oxygen transfer limitations etc.), if it is to be operated bubble-free.

A low shear and bubble-free alternative to the use of microcarriers and macro-porous carriers for the propagation of anchorage-dependent cells lies in the application of Packed Bed Reactors (PBR). However, it has been demonstrated that this reactor type can also be employed as an efficient means of propagating suspension cells (demonstrated with hybridoma cell lines, see below).

It also appears that the high local and overall cell densities (based on the cell aggregates and low total reactor volume, resp.), a feature of this reactor, enable a significant reduction in the amount of protein required for serum-free cultures. For example, it has been possible to propagate some hybridoma lines continuously (over several weeks) using protein-free medium alone; the same lines could not be maintained without the addition of transferrin and insulin in a stirred (low shear) spinner flask or 15 liter reactor under otherwise comparable culture conditions (Bliem R, Varecka R, Oakley R, publication in preparation; Fig. 4).

This reactor type comprises a culture vessel packed with inert particles (glass, polyurethane, ceramic, etc.), usually solid or porous spheres in the range of 2 to 5 mm in diameter [13–16]. This packed vessel is incorporated within a medium recirculation

Table 1. Selection criteria for airlift reactors

Process requirements	Reactor features (current status)
Volumetric capacity	preferable: > 10 l reactor volume; largest units: 800 to 2000 l reactor volume (Bender Austria; Celltech, UK);
Typical cell count per reactor unit	1 E10 to 8 E12[a]
Typical cell density	2 to 4 E6 cells per ml
Number of parallel products	few
Mode of operation	batch or continuous (perfusion difficult);
Typical length of run	weeks to months;
Process control	homogeneous cell suspension; direct on-line monitoring and control of pH, pO_2 etc.
Cell line stability	generally required to be > 50 generations;
Product type	cell mass production; secreted or cell-associated product; virus production;
Medium	protein and anti-foam additions usually required;
Scale-up	simple design and engineering as unit process system; economic automation; good economy of scale;
Operational constraints	bubble-associated damage to microcarrier-based cultures likely;
Operating benefits	no pumps; no oxygenator; no moving parts;
Maintenance	simple maintenance; steam sterilization;

[a] E represents the computer notation for "10 to the power of"

loop, connecting the reactor with an oxygenator. Cells initially occupy the surface of the packing. As the cells multiply they begin to fill the void space between the particles, the extent of which depends on the structural features of the packing. As the cells fill the void space, and over extended culture periods (months), the reactor becomes prone to channelling and gradient formation; these effects are exacerbated on scale-up. A new reactor configuration alleviates these effects (Fig. 5).

Scale-up of this reactor configuration is simple and may be achieved on the basis of linear velocity alone, provided that the packing is uniform and has been well charac-

Table 2. Selection criteria for fluidized and packed bed reactors

Process requirements	Reactor features (current status)	
	Packed bed reactors	Fluidized bed reactors
Volumetric capacity (continuous perfusion)	up to approximately 300 l per reactor per day ($\sim$ 1000–2000 l batch equivalent; Bio-Response, USA; Verax, USA; IAM, Austria)	
Number of parallel products	several possible	
Typical cell count per reactor unit (current status)	up to 2 E12 cells	
Typical cell density (based on bulk working volume)	5 E6 to 5 E7 cells per ml (aggregate densities up to 5 E8 cells per ml)	
Mode of operation	continuous operation over extended periods most economic, although batch possible;	
Process control	indirect measure of cell number and condition only	macroporous carriers can be sampled but representative cell measurements problematical
	pH, pO_2 etc. measured on-line in recirculation loop;	
Cell stability	cell line should be stable over $>$ 50 generations;	
Product type	preferably secreted products; virus production possible;	
Medium	use of protein-free media apparently facilitated;	theoretically high medium utilization due to high cell densities and mixed suspension;
Scale-up	economy improves on scale-up; simple scale-up design;	
Operational constraints	oxygenators and pumps can present engineering problems; complicated to use as seed stock reactors; potential diffusion limitations;	
	no direct measurements of cell mass;	cost and design of carrier matrices; representative cell measurements difficult;
	handling effort of carrier matrix	
Operating benefits	cell retention; low shear stress; reusable packing; low inoculation densities (possible (0.01 of harvest count);	homogeneous aggregate suspension (facilitates control); cell retention; low shear stress;
Maintenance	packing and carriers require handling effort; steam sterilization;	

terized, both under cell free and under culture conditions. This system therefore also offers many of the features desirable in cell culture, i.e. high cell retention and perfusion capabilities, simple scale-up and bubble-free operation, in addition to the potential medium benefits enabled by the high cell densities. The ability to obtain a direct and representative measure of the cell densities and viability remains to be developed. The recirculation pumps and the oxygenators are also critical elements of this system (and have therefore been the focus of a development program at Bio-Response).

In principle membrane reactors, such as hollow fiber (see below) or flat sheet membrane reactors [17] differ from the packed bed reactor in that the packing consists of a permeable membrane, seperating the bulk medium flow from the cell mass. The medium may be perfused (convective flow) through the cell mass and collected on the cell side, or exchanged (largely) by diffusion across the membrane. This type of reactor is very well suited to the production of moderate quantities of several different antibodies simultaneously; a typical reactor unit may contain between 1 E10 and 1 E11 (E denotes "to the power of 10") cells, although the viability may vary considerably with operating conditions [18–21]. Whereas suspension cells perform well in this reactor type, anchorage-dependent cells are prone to clog the reactor membrane and introduce significant diffusion limitations. The membrane material should also be tested for product binding properties.

Table 3. Selection criteria for hollow fiber and stirred tank reactors

Process requirements	Reactor features (current status)	
	Hollow fiber reactors	Stirred tank reactors
Volumetric capacity	4 to 8 l per unit per d (per ~ 50 ml extracapillary space and ~ 0.5 m² extracapillary extracapillary surface)	Preferable: > 5 l per reactor per d; Largest reactors: 5000 to 10000 l reactor volume (Wellcome, UK; Genentech, USA);
	up to 200 units per facility practical (Xoma, USA; Bio-Response, USA)	
Typical cell count per reactor unit	2 E9 to 5 E10 (flat-membrane reactors up to 5 E11)	1 E9 to 5 E13
Typical cell density	1 E7 to 5 E8 cells per ml;	2 to 4 E6 cells per ml; up to 3 E7 cells per ml possible with cell retention;
Number of parallel products	several products (flexible production system)	few products
Mode of operation	continuous perfusion preferable	batch or continuous; (perfusion possible)
Typical length of run	months	weeks (months)
Process control	indirect measure of cell density and condition; heterogeneous cell mass;	direct monitoring and control of homogeneous cell mass;
Cell stability	cell line stability dependent on scale and duration of culture (generally > 50 generations) (both systems);	

Table 3. (Continued)

Process requirements	Reactor features (current status)	
	Hollow fiber reactors	Stirred tank reactors
Product type	secreted product;	cell mass production; secreted or cell-associated product; virus production;
Medium	high cell densities facilitate use of serum-free media;	optimum medium supply to cells in single-cell suspension;
Scale-up	linear increase in cost; good predictability if increase in capacity by multiple units; (reactor scale-up difficult); low initial capital costs	economy improves on scale-up; scale-up as unit process; economic automation; low running costs (compared for example to HFRs);
Operational constraints	control limitations; increasing number of units on scale-up; impractical as seed stock reactors; diffusion limitations;	using direct sparging: bubble-associated cell-damage; foaming; (bubble-free oxygenation now possible for industrial scales);
Operating benefits	bubble-free oxygenation; cell-free supernatant (simple cell retention); on-line concentration possible; low space requirements for small product quantities;	good in-process control possible;
Maintenance	disposable units; labor intensive routine; sterilization: irradiation, ethylene oxide (rarely steam);	sterilization: steam;

The relatively simple operating requirements of hollow fiber reactors are a distinct benefit, although they rely on special techniques and skills. The reactor units may be obtained at a cost at which their disposability is still economic. The reactors may be operated in such away as to yield cell free and product-concentrated supernatant, which translates into significantly lower down-stream processing costs.

Scale-up is currently only possible by multiplying reactor units. Although this limits their economy of scale, their small size and simple design facilitates easy handling, and the simultaneous production of several products within the same facility (Bio-Response Inc. uses Hollow fiber reactors routinely for the production of up to 100 g quantities of several monoclonal antibodies simultaneously).

Finally, in Tubular Reactors the cells are propagated whilst slowly flowing along the length of a tube or along a series of connected reactors. A reactor of this type is currently being developed for industrial application [8, 22].

4 Propagation Mode : Batch vs. Continuous

A determination of the propagation mode, batch or continuous, should be based on a careful analysis of the product formation characteristics, capital and operating expenses, and the required level of process control, raw material and product quality.

A typical batch culture lasts between 5 and 15 days, depending on the inoculation density and the culture growth characteristics. After an initial adaptation period (lag phase), the growth rate of the culture increases, until nutrient depletion and waste product formation limit continued growth and subsequently lead to culture degeneration and lysis.

Some two thirds of the duration of a batch process comprise culture growth. Therefore this form of propagation is an option for products which are formed preferentially during culture growth (growth-associated product formation), or in which the product is formed both during growth and non-growth (lag or stationary) phases (constitutive product formation). However, if the product is formed constitutively, as appears to be the case with most proteins of current industrial interest, including monoclonal antibodies, continuous production systems offer economic advantages; the underlying principles are well established for microbial and chemical processes [23–25] and equally apply to animal cell culture [8]; Bender GmbH and Bio-Response, unpublished information).

One obvious advantage of continuous culture therefore lies in the savings of repeated growth cycles. Continuous cultures are generally operated well below the maximum growth rate, thereby reducing the medium wasted on biomass production (this is of course of little value if the desired product is biomass associated, e.g. membrane antigens, or involves lytic virus production).

Operating expenses are also comparatively lower for continuous culture as a result

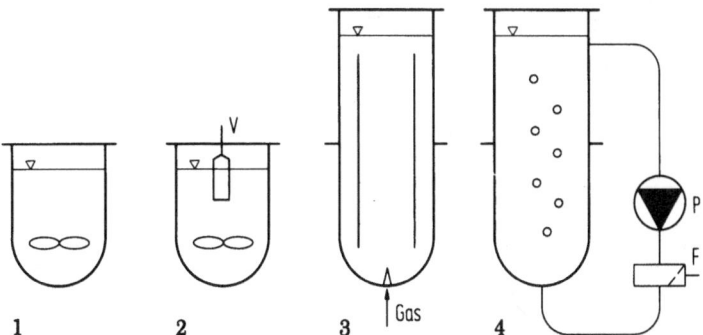

Fig. 6. Exchangeable reactor configurations (Unit Process Modules) for continuous process operations at IAM
1 STR, standard microbial reactor configuration with axial flow impeller; Hl/Dt $\sim$ 3; *2* STR, as 1., with vibro-cage (*V*) for oxygenation and cell (microcarrier) retention; *3* Airlift, with or without vibro-cage; Hl/Dt $\sim$ 6; *4* Airlift configuration as Fluidized Bed with on-line cross-flow filter module (*F*); (*Hl* ... Liquid Height   *Dt* ... Tank Diameter
P ... Recirculation Pump)

of more efficient labour utilization, reduced Mean Down-Rime and the associated turnaround costs.

There is little overall difference in capital costs between the two modes of operation (relative to the total plant costs). The holding capacity of peripheral vessels is generally lower for continuous culture as the bulk liquid flow is quasi continuous (the medium is prepared more frequently in smaller batches), but on the other hand requires more equipment and piping. Conversely, peripheral vessels of similar size to those required for batch culture, potentially offer a comparatively larger production capacity.

The reduced vessel size also enables greater flexibility. This is of significance because there are few production plants that are yet dedicated to a single product or cell line; flexibility therefore translates into greater plant economy (see illustrations of modular reactor configurations used at IAM and Bio-Responses, see Figs. 6 and 7).

The relative performance stability of continuous culture facilitates process control, process and product assurance. The dynamic transients, which characterize batch culture per se, complicate process control and assurance.

The product concentrations are typically (although not necessarily) higher in batch culture, which has a significant beneficial impact on the processing costs. On the other

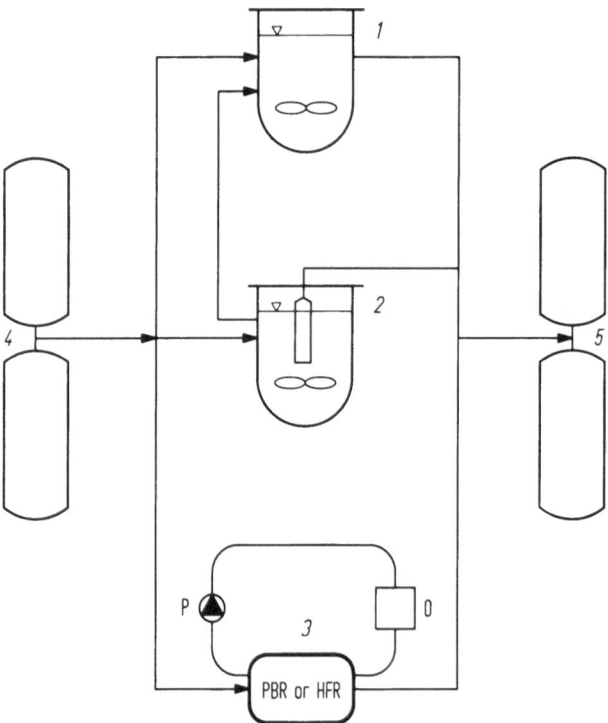

Fig. 7. Conceptual layout of reactor modules used for continuous process operations at bio-response *1* Batch Reactors recycled in series to enable semicontinuous operation of lytic virus process (e.g. Baculovirus); *2* STR with cell retention for continuous product formation or cell mass production (e.g. growth factor or surface antigen); *3* Packed Bed Reactor (*PBR*) or Hollow Fiber Reactor (*HFR*) for continuous (secreted) product formation (anchorage dependent cells and Hybridomas, respectively); *4* Medium Holding Tanks; *5* Harvest Collection Tanks;

hand cellular by-products, such as proteases and DNA also accumulate; in some processes these products of cell lysis need to be monitored. This aspect of batch culture is not only of technical interest, in terms of quality assurance, but is also gaining interest for economic, and for regulatory reasons. Recent trends in the industry and within the regulatory agencies indicate that DNA is not only of concern in the final product, but should also be monitored throughout the process, suggesting that the DNA concentration should already be minimized in the raw cultured supernatant, that is prior to purification.

It is fair to say that the physiological patterns of animal cells as well as the process economics are forcing a general trend towards using continuous, and preferably perfusion systems (in which cells are retained within the reactor). This is not only a theoretical consideration but is increasingly being demonstrated throughout the industry and academic institutions alike. For example, the use of smaller reactor vessels in continuous culture has enabled a modular reactor and vessel configuration in the design of a 4 million dollar pilot plant at the IAM in Vienna. The modular process units (Fig. 6) permit a much higher level of operational flexibility than is otherwise possible with batch systems. At Bio-Response Inc. a similar approach (Fig. 7), based on a combination of stirred perfusion reactors, hollow fiber and packed bed reactors has made it possible to economically propagate a wide spectrum of cells lines simultaneously, within the same plant employing minimum size vessel, both for the reactors and the periphery. During the year 1987/88 this plant processed some 800000 liters of culture supernatant.

5 Perfusion and Cell Retention

To operate a reactor in "perfusion" mode denotes a system in which cells are retained within the culture vessel, but in which the medium is continuously renewed, i.e. perfused. Although a perfusion culture may be operated in batch mode, i.e. with a medium reservoir in an external loop [14], perfusion systems are generally operated continuously [16, 19].

Cells may be retained within the reactor by immobilisation (e.g. in macro-porous carriers in fluidized bed systems, or within the matrix of packed bed reactors or by membrane retention in hollow fiber reactors). In recent years spin and vibrating filter

Table 4. Relative comparison of some economic features of different modes of operation (high + + +, medium + +, low +)

	Batch	Semi-continuous & fed-batch	Continuous & perfusion
Operating expense	+ + +	+ +	+
Capital expense	+ + +	+ + +	+ + +
Labor	+ + +	+ + +	+
Process control	+	+ +	+ + +
Raw material quality	+	+ +	+ + +
Product concentration	+ + +	+ +	+
Required skill level	+ +	+ +	+ + +

devices have been developed for STRs to retain microcarriers [26, 27] and cell aggregates [28]. The use of a vibrating movement has facilitated the development of a device which functions both to retain microcarriers and cell aggregates as well as functioning as an efficient oxygenator [22].

Similar devices used for single cell suspensions typically clog after about a week, or do not possess sufficient retention efficiency to maintain cell densities beyond 8 E6 cells per ml. Investigations into the mechanisms of filter clogging have led to the development of a device which so far has enabled hybridoma cultures to be maintained at or above 10 E7 cells per ml for several weeks, amounting to over 90 reactor volumes perfused through the system before filter clogging set in (unpublished).

6 Process Definition and Objectives

A clear description of the process objectives and basic characteristics are an essential prerequisite for the efficient selection and development of a process system. Some amount of development will always be required for a new application, even if the reactor system itself has already been well characterized. The goal of such a development program is to define the specific process requirements and to provide solutions

Table 5. Cell identification

Species
Cell line age, preferably in generations;
Condition of source material (tissue or cells of origin):
   normal or tumor.
If transformed, source of transformation: viral, chemical, irradiation, oncogene (detail);
Tissue (or cell type) of origin;
Characteristic cell markers, karyotype, isozyme pattern,
   population homogeneity (FACS analysis);
Virus and Mycoplasma contamination (tests for virus permissivity can be useful);

Table 6. Culture characteristics

Morphological characteristics;
Subcloning efficiency and clonal stability;
   Percentage of producer cells in population and stability;
Substrate adhesion characteristics and requirements:
   predominantly suspension or anchorage growth type?
   if anchorage type, has it been propagated by microcarriers, macrocarriers or in aggregate form?
Routine expansion techniques: flasks, rollerbottles, spinners, shaker flasks, mass culture reactors
   (detail);
Typical growth rate or culture doubling time obtained in above culture techniques (for batch
   culture give average doubling time);
Number of cell doublings for culture expansion from cell bank to production process; split
   frequency in batch culture;
Inoculation densities; Typical maximum cell densities;
Cell freezing and Master Cell Bank (MCB):
   cell densities and generations from cloning;
   cryoprotective agent and medium;
   plating efficiency of frozen stock;

(parameters) which will form the basis for the final process design. Tables 5 through 10 list examples of some typical information necessary to determine the process requirements (only the items pertaining to the cell process are detailed; purification and regulatory issues are beyond the scope of this discussion).

Table 7. Culture medium

Composition of the media for growth and production phases;
Quality specifications for: endotoxins, trace metals, microbial, and
  particularly viral contaminants, aminoacid profile and salt composition
  of protein hydrolyzates, etc.;
Preparation protocol (solubilization, filtration etc.);
Component stability (4 °C, 37 °C);
Induction/selective/stabilizing agents for product
  formation/culture stability;

Table 8. Product formation

Product quality (heterogeneity, contamination) and constancy;
Product stability at 37 °C, and at 4 °C (for all process steps); Freeze-thaw
  stability;
Product affinity to typical materials used (e.g. polypropylene);
Product formation/secretion rate under growth and non-growth conditions
  (typical ranges); product induction, cell lysis associated etc.;
Product affected by presence of serum, cellular proteases etc.;
Product purification and final product formulation protocol;

Final product specifications

Process and product assays

Regulatory requirements

Table 9. General process descriptors (examples)

Product characteristics (material, functional and economic);
Basic reactor design and mode of operation;
Process control operating specification;
Required annual production capacity and production rate
  (acceptable range) and production frequency;
Required minimum product titers;
Bulk liquid handling capacities;
Total duration of a single process run;
Acceptable process down-time;
Duration of purification process;
Allowable storage period of culture supernatant;
Number of products and cell lines to be produced within the same system;
Extent of fresh medium quarantine for QA/QC;
Electricity, steam, water and gas usage;
Facility and space requirements/availability;
Scale specific features related to medium preparation, product purification or waste treatment;
Staff requirements;

Table 10. Reactor parameters (examples)

Method of liquid agitation (impellers, pumps for recycle system);
Critical process parameters (mass transfer, power input, liquid
    velocity, etc. and associated operating parameters);
Maximum oxygen transfer rate for gas-liquid mass transfer required (K_L, K_La, OTR as appropriate);
Liquid-solid (liquid-cell) mass transfer characteristics
    (Sherwood Number, Thiele Modulus, Dispersivity, maximum cell depth, etc., as appropriate
    for cell multilayer based cultures);
Volume of agitated liquid;
Liquid dispersion characteristics (e.g. Peclet Number, if appropriate);

The information listed in Tables 5–8 will enable a first description of the process intended for development and scale-up. The following two tables (9, 10) list further information requirements for the design and construction of a new production facility.

7 Selection and Development of Reactor System

In the past cell culture reactors were adopted, rather than selected, on the basis of available expertise, equipment, consulting and other services, etc. Until the mid 1970s traditional microbial technology provided the main source of knowledge and equipment for industrial processes, almost exclusively in the form of stirred tank operations.

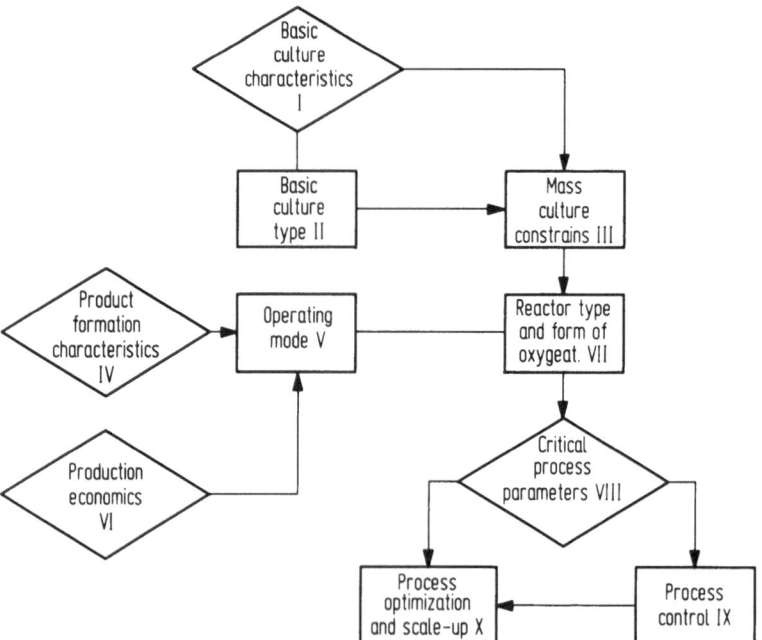

Fig. 8. Overview of relationship between process characteristics and reactor selection and development

By 1980 a small number of unit process reactors had been developed specifically for mass cell culture: the airlift [6] hollow fiber units [18], Glass Bead Packed Bed Reactor [14], and Rotating (Disk) Trickle Bed Reactors (Connaught Labs., GD Searle during the late 1970s — no publications available).

Today we have a relatively large number and variety of practical reactor systems, which are available from almost as many vendors. Unfortunately useful comparable engineering data, which could serve as an objective means of assessing the various reactors are rare.

Each reactor system has a unique set of characteristics (including economic), which allow for a rational selection of an appropriate reactor in accordance with the process and business requirements.

The selection of a reactor type will be determined by the:

— culture characteristics of the cell (e.g. suspension-type; sparging or bubble-free oxygenation etc.)

— product formation characteristics (e.g. constitutive or growth-associated production)

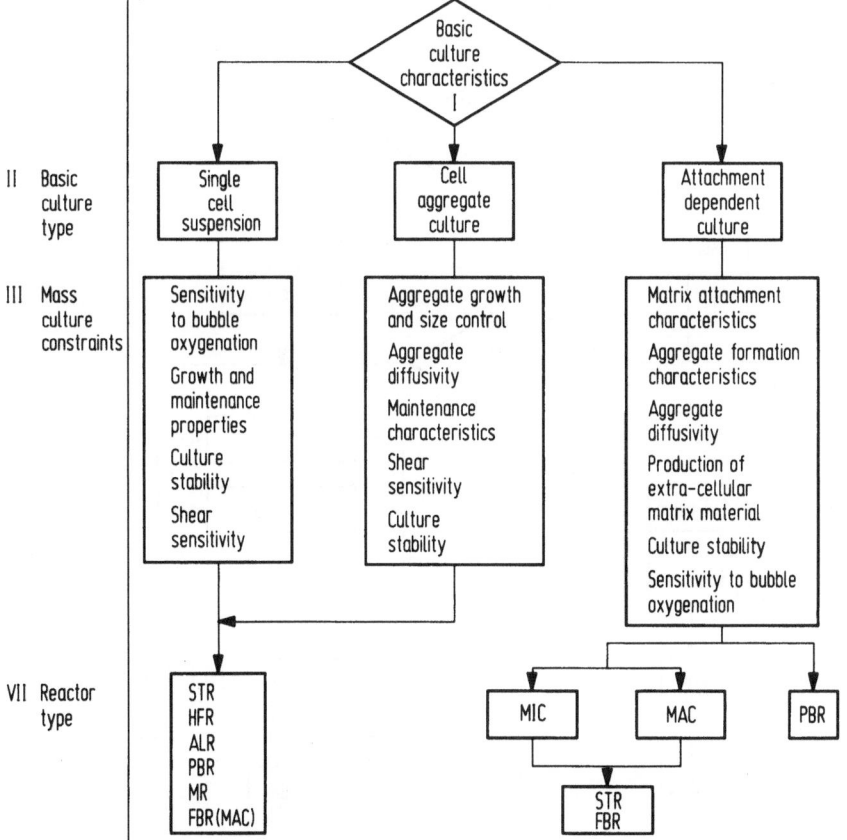

Fig. 9. Relationship between basic culture characteristics and reactor selection (strategy)

— techno-economic considerations (e.g. required production capacity; single or multi-product operation etc.)

Figure 8 illustrates how these aspects relate to the selection of process features. The basic culture characteristics (Fig. 9) will determine the propagation type (anchorage dependent, aggregate or single cell suspension type, etc.), as well as the potential culture constraints (e.g. complicating growth or maintenance characteristics). Figure 9 illustrates how the culture properties have an impact on the selection of a reactor type.

Figure 10 shows the product formation characteristics affect the operating mode, which in turn determines the reactor type. For example, in the case of a commodity product (e.g. low value diagnostic) with a low shelf life will be mainly concerned with minimizing medium and capital costs, using a system which can respond quickly to changing demands continuously or in frequent batches as required. On the other hand, with a high value therapeutic, more emphasis would be placed on aspects such as process control, for technical as well as regulatory reasons, and on scalability (so that the process can be frozen as early as possible). Obviously such a simple delineation is not always possible.

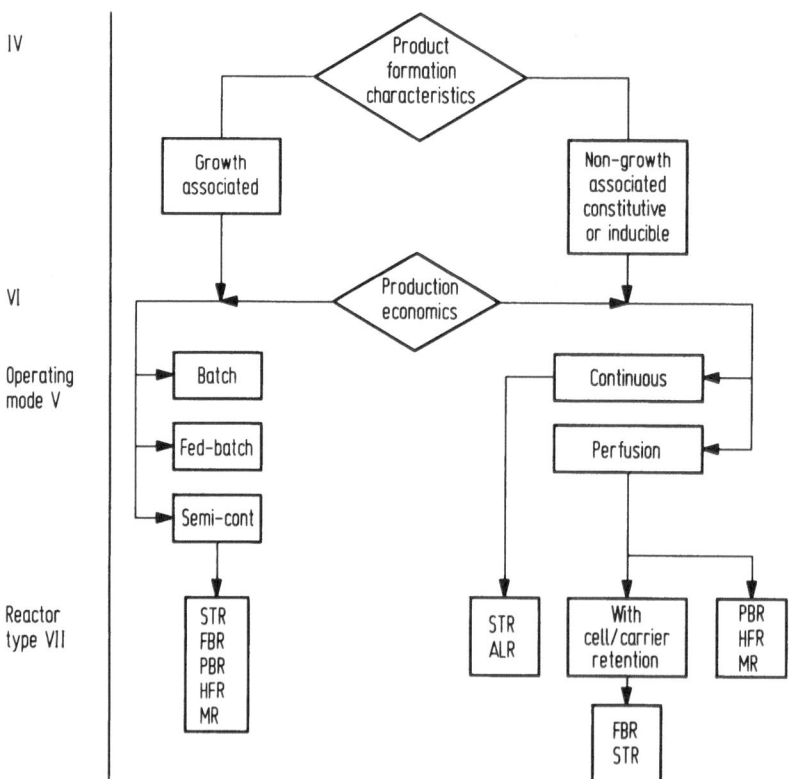

Fig. 10. Relationships between product formation characteristics and reactor selection (strategy)

Figure 11 presents an outline of the possible strategies for the design and scale-up engineering of the process. Table 11 lists an "ideal" combination of process conditions; the design and scale-up procedure will depend on the extent to which any of these critically limit the process objectives. Generally, scale-up of animal cell culture processes are reducable to a balance function between (non-diffusion limited) gas

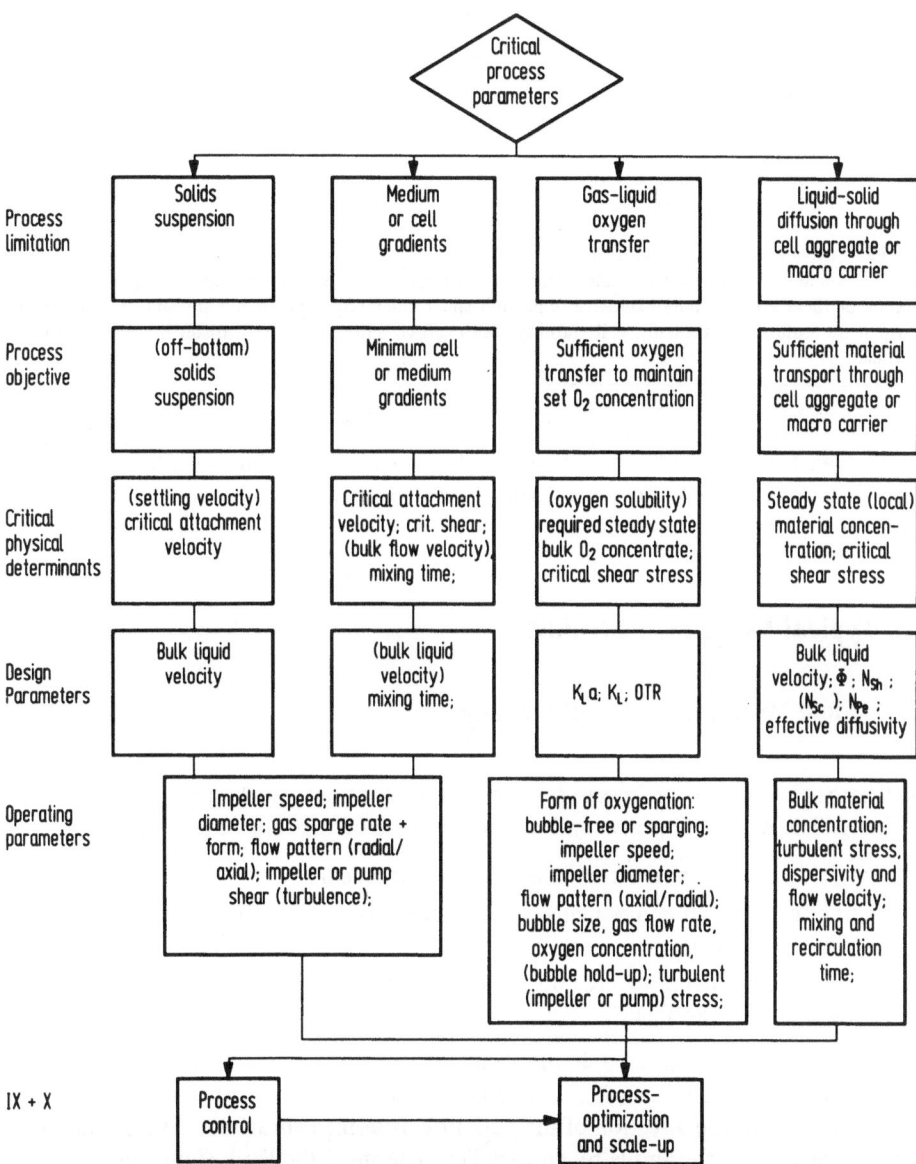

Fig. 11. Aspects of reactor optimization for cell aggregates, oxygen transfer, medium blending and solids suspension (these are ideal delineations)

Table 11. The Ideal Reactor

An ideal cell culture reactor may be described by the following requirements:
1 — homogeneous cell and medium distribution
2 — no mass transfer rates limitations
3 — no damaging effects of shear stress
4 — no damaging bubble effects
5 — no foam formation
6 — cell densities $> 5 \, E7 \, ml^{-1}$
7 — temperature control $37 \, +/-0.1$ °C (most mammalian cells)
8 — no local hot-spots above 37.5 °C (most mammalian cells)
9 — variable range of working volume to $\pm 1\%$ total capacity
10 — linear scale-up on the basis of one easy-to-determine and operable process parameter
11 — minimum space requirements
12 — minimum maintenance requirements
13 — inert reactor materials
14 — simple design configuration
15 — inexpensive construction
16 — proven construction techniques
17 — proven operating procedures
It is unlikely that there is a single optimum combination for all these parameters, particularly as some of them are mutually exclusive (e.g. high shear rates required for mixing and mass transfer, whilst we require low shear for cell viability; both cell viability and mass transfer determine process efficiency)

mass transfer and (minimum) shear or bubble-associated cell damage. Shear sensitivity is a function of inherent cell properties, recent propagation history, medium reactor design and operating conditions.

8 Typical Design Constraints

8.1 Fluid Dynamics and Cell Damage

All large cell culture processes require liquid agitation, which will result in liquid mixing. The mixing objective will be one or more of the following:

maintaining cells or carrier matrices in liquid suspension
Blending the medium components so as to minimize any gradients (bulk mixing, as opposed to molecular dispersion)
dispersing medium components on the molecular level (gas-liquid, liquid-liquid and liquid-solid mass transfer)

Any liquid mixing process is the result of velocity gradients within the agitated liquid. In its simplest form, i.e. under laminar flow conditions, the velocity gradient perpendicular to the direction of the bulk flow is termed shear rate. The liquid shear rates generated, for example by an impeller, determine the level of mixing.

Although shear is therefore a basic operational requirement, it may also be damaging to animal cells. Fortunately, it is usually possible through empirical development to operate within shear limits that permit the process objective to be achieved (see above), without causing significant cell damage. Shear damage is attributed to the product of shear rate and liquid viscosity, i.e. shear stress. Accurate, and above all relevant shear stress measurements, which reflect practical working conditions (as a function of shear, turbulence, eddy size, power input, viscosity, flow patterns etc.), are difficult to obtain, particularly in large scale stirred tank reactors. This difficulty is compounded by the fact that shear patterns in stirred tank reactors become increasingly complex on scale-up. In order to estimate the extent to which such shear damage may occur, we must be able to characterize, in qualitative and quantitative terms, the nature of the hydrodynamic conditions generated, and relate these terms to reactor operations. Unfortunately such a generally applicable concept, comprising practical engineering parameters which would enable reliable and direct comparisons between different reactor systems (e.g. between stirred tank and packed bed reactors), has yet to be established).

The shear sensitivity of animal cells may be taken into account in the design and scale-up procedure for a reactor. For example, stirred reactors may be designed and scaled towards maintaining a critical oxygen transfer rate, or, if it is possible to satisfy the oxygen requirements independent of impeller speed, then it may be sufficient to design and scale with the objective of merely maintaining a uniform medium dispersion. Lastly, if medium gradients also do not affect the process performance, then the process could be designed to maintain the cells in suspension (the authors are aware of only one example in which this has been put into industrial practice). The maximum and average shear stress levels decline in the expected order, so that designing for suspension would be most desirable. In practice, the power which is required to prevent cells from settling out and "sticking" in pockets of the vessel is generally as high as that required for blending operations. Generally stirred tank reactors (BTRs) will exert the highest levels of maximum shear (disregarding the effects of sparging). In fluidized bed reactors the larger proportion of cells is protected within the macroporous carriers, whilst in packed bed reactors the fluid velocities are so low that they exert no significant damage. Under usual operating conditions membrane reactors are practically shear free.

8.2 Bubble-Associated Cell Damage

It is a common observation that direct gas sparging of the culture medium may be deleterious to cells grown in free suspension (as single cells or cell aggregates), but particularly to cells grown on microcarriers. It is for this reason that air-lift systems have not been generally applicable to anchorage dependent cells grown on microcarriers. Sensitivity to gas sparging depends on the cell type, the bubble size, the bubble frequency, the superficial gas velocity and the form of cell propagation [29]. Although calculations indicate rising bubbles may exert potentially damaging levels of local shear stress [30, 31], it appears that velocity gradients in the fluid film adjacent to the gas-liquid interface are not the dominating mechanism in bubble induced cell damage. It has been demonstrated that significant cell damage occurs in the process of bubble disengagement [29].

Although the underlying reason is not understood, the addition of surface active agents such as Pluronic, Silicon Oil or Pluriol may improve cell viability under conditions of sparging. Similar results have been achieved by the addition of viscosity enhancing agents such as methylcellulose or polyvinyl pyrrolidone [32].

Clearly, we need a mechanistic theory which relates the dynamic effects of surface tension forces, surface and bulk viscosity and hydrodynamic forces to advance on the empirical observations of cell damage due to liquid agitation and gas sparging.

8.3 Oxygenation

8.3.1 Oxygen Uptake Rates (OTR)

The oxygen uptake rates (OTR) of most cells fall into the range of 0.10–1.5×10^{-16} mol s^{-1} O_2 per cell [8, 16, 22, 33]. One cm^3 of a packed cell mass contains about 5×10^8 cells (cells of about 15 μ in diameter). The OTR of a cell culture of maximum density would therefore be in the region of 0.1–2×10^{-3} cm^3 ml^{-1} s^{-1} O_2 per volume suspension. The $K_L a$ values for various oxygenation systems range up to 7×10^{-3} s^{-1}. Therefore, were it not for other problems associated with gassing (see below), gas-liquid mass transfer and the oxygen supply of single cell suspension cultures should not present limitations as $k < K_L a$ (k is the O_2 consumption rate and $K_L a$ is the overall oxygen transfer coefficient).

Similarly, the rate of mass diffusion through a cell aggregate should at least equal to the rate of oxygen consumption. Based on a diffusivity of 1 to 3×10^{-5} cm^2 s^{-1}, the oxygen solution rate is 1.79×10^{-2} ml ml^{-1} bar^{-1} [34, 35] and the wide range of oxygen consumption rates it may be calculated that oxygen diffusion is likely to be significantly limited (less than 10% air saturation) at a cell, or matrix depth of 500 to 1000 microns, although calculations range to as low as 100 microns [36]. This calculation applies to hollow fiber, membrane, packed bed and fluidized bed reactors.

8.3.2 Gas Sparging

The following points concerning gas sparging are most relevant to STRs and ALRs, although gas sparging may of course also be used in connection with PBRs, FBRs or HFRs.

It is usually necessary to maintain a bulk liquid flow of 3 to 5 cm s^{-1} in order to prevent cells from settling out in vessel pockets; of course this is much higher than the settling velocity of the cells, which is all that would be required in an ideal reactor (to maintain cells in suspension).

As airlift reactors rely on gas sparging both for liquid mixing and gas transfer one might expect this type of reactor to be most sensitive to the constraints of bubble aeration. We usually operate our reactors in such a way so as to achieve a bubble diameter of 1 to 4 mm, with a superficial gas velocity of 0.02 to 0.07 cm s^{-1} and an absolute terminal bubble velocity in the region of 10–30 cm s^{-1} (employing gas flow rates in the region of 1 VVh). For STRs, typical gas flow rates (for oxygenation) using air are in the region of 0.1 to 1 vvh. The addition of an antifoam agent is usually needed (or other means of foam inhibition), particularly upon scale up as the superficial velocity increases (the gas flow is increased in proportion to the cube of the scale-up ratio, whereas the degassing area increases only by the square). Larger STRs are therefore more prone to severe foaming.

On the other hand the introduction of an antifoam agent at the scale-up stage may be undesirable due to the problems and effort associated with changes to regulatory license applications and purification procedures. Surfactants (including antifoams) may affect the purification process, however the significance of this effect will vary from product to product.

Although small bubbles have been shown to increase cell damage, microsparging has been successfully applied to the growth of hybridoma cultures, particularly in combination with oxygen enriched air and even pure oxygen.

8.3.3 Bubble-free Oxygenation

It is evident from the above that bubble free oxygenation would be preferable. This has been approached in several ways: head space gassing, gassing through gas-permeable membranes, or sparging within a wire cage.

The application of head space gassing is limited to laboratory vessels. Silicone membranes on the other hand have been successfully applied to industrial systems, although there is much "know how" involved, which is not available in the literature. Wire cage oxygenators have also been shown to be very efficient and practical (see Table 12, vibro-cage). In principle gassing is restricted to an area within a wire gauze cage while the gauze prevents foam from escaping into the bulk liquid. In order to promote mass transfer the cage is agitated [27, 37].

Table 12. Comparison of oxygen transfer rates of different oxygenation systems (using pure oxygen, 20 % air saturation assumed)

System	Oxygenation transfer $mg\ m^{-2}\ h^{-1}$
head space diffusion	2 880 (270)[a]
microsparging	
(ceramic sparger, 0.2 μ)	2 E6 (5340)[b]
permeation across silicone tubes	
0.2 mm thickness, 0.7 bar	26 000 (5340)
0.05 mm thickness, 0.7 bar	57 000 (10 800)
kinetic diffusion across	
vibro-cage	48 000 (4920)

[a] Values in parentheses: absolute OTR in a 35 l reactor in $mg\ h^{-1}$.
[b] Data from W. Scheirer

9 Extent of Evaluation

Industrial process systems must enable the process objectives to be achieved in a simple, safe and reproducible manner. Once the process hardware, design and operating conditions have been determined and the registration procedures initiated, the introduction of significant modifications will be discouraged by the ensuing regulatory requirements, at least until the manufacturing process is well underway. The selection of a process system must therefore be based on a careful experimental investigation of its performance parameters. An assurance of the stability and the reproducibility of the process is possibly the most important test criterion.

The experimental evaluation of a process intended for industrial application should be extended to the point, where a minimum set of parameters is made available to specify basic closed-loop control (pH, pO_2, temperature pressure, gas flow, liquid flow, level pC_2, temperature, pressure, gas flow, liquid flow, level indicators etc.).

As industrial cell culture processes are now rarely operated in batch mode alone, provisions should also be made for fed-batch, or preferably continuous operation. The ensuing necessity in developing an algorithm for the controlled addition of the medium requires an understanding over and above that necessary for batch processes.

In order to assess the stability, robustness and reproducibility of a system, several runs will be necessary. The evaluation must be extended to include considerations such as space or facility constraints, availability of skilled personnel, reliable services and supply of process materials (e.g. the risk involved in an all too strong dependence on a single supplier for process materials). The selection of a process system will need to take into account the likely applications for years to come and the extent to which the system should allow for operational flexibility. "Dedicated" plants do not always remain dedicated to a single product or producer line, as we have learnt from the fermentation industry.

These are of course both matters of process and business strategy, and illustrate how decisions concerning process design go beyond just the technology alone. The industrial development of a process is likely to be initiated by the business prospects and must include a business perspective throughout the development program.

10 The Need for Standard Test Cell Lines

The imagination of cell culturists and possibly an all too subjective approach to reactor design has brought about a bewildering spectrum of reactor configurations. The current trend of investigations indicates that the apparent variety of cell types may be classified into a few categories with respect to process requirements, which will in turn be satisfied by a small group of principle reactor types.

In order to finally permit a wider and systematic comparison of the general performance characteristics of the various reactors, we, the cell technologists, should agree to adopt a small number of relevant reference cell lines made available through the public cell banks. Such cell lines (possibly up to 6 lines) should possess culture and product formation characteristics which are relevant to industrial situations. A more detailed analysis and widespread discussion would be necessary for this subject.

The need for such standard test cell lines is becoming increasingly pertinent as more and more organizations turn to animal cell processes for new product lines (as opposed to single products). The efficient selection and evaluation of a process system, with a view to a wider application, requires that one is able to distinguish between the general performance characteristics of a reactor and the inherent properties of the cultured cell line.

11 Future Prospects

It is conceivable that with an increased understanding of the relationship between the structure and the biological function of proteins, we will be in a position to construct improved cell-based systems by means of protein engineering and hybridization techniques, which will replace the current tissue derived cell types.

At least for the near future, however, we will have to rely on tissue-derived animal cells as a production vehicle in biotechnology and must contend with the complications of process design and development, which arise from the cells' mechanical and and physiological fragility. Nevertheless, whatever the nature of the new cell constructs, they are still likely to present process characteristics akin to the current cell types; the principle issues of mass propagation and maintenance of fragile cells with complex requirements will remain a challenge.

12 Abbreviations

ALR	Airlift Reactor
FBR	Fluidized Bed Reactor
HFR	Hollow Fiber Reactor
$K_L a$	Overall gas liquid transfer coefficient
K_L	Area specific gas transfer coefficient
Mac	Macroporous carriers
Mic	Microcarriers
OTR	Oxygen Transfer Rate
Nsh	Sherwood Number
Nsc	Schmidt Number
Npe	Peclet Number
Phi	Thiele Modulus
PBR	Packed Bed Reactor
STR	Stirred Tank Reactor
MR	Membrane Reactor
VVh	Gas Flow Rate in volumes gas per reactor volume per hour

13 References

1. Beale AJ (1979) Adv. Exp. Biol. Med. 118: 83
2. Hopps HE, Petricciani JC (eds) (1985) Abnormal cells, new products and risk, Tissue Culture Association, Gaithersburg
3. Lubiniecki AS, May LH (1985) Develop. biol. Standard. 60: 141
4. Baker P, Knoblock K, Noll L., Wyatt D, Lydersen B (1985) Develop. biol. Standard. 60: 63
5. Pay TW, Boge A, Menard FJ, Radlett PJ (1985) Develop. biol. Standard. 60: 171
6. Katinger HW, Scheirer W, Kromer E (1978) Germ. Chem. Eng. (Engl. Transl.) 2: 31
7. Birch JR, Thompson PW, Lambert K, Boraston R (1984) The large scale cultivation of hybridoma cells producing monoclonal antibodies. ACS Annual Meeting 1984, Philadelphia (Abstract)
8. Katinger H (1987) Develop. biol. Standard. 66: 195
9. Porter BF, Schurr GA, Wells DF, Semrau KT (1984) Fluidized-bed systems. In: Perry RH, Green D (eds) Perry's Chemical Engineering Handbook, chap 20, p 58

26		R. Bliem, K. Konopitzky, H. Katinger

10. Andrews GF, Przezdziecki J (1986) Biotech. Bioeng. XXVIII: 802
11. Shirai Y, Hashimoto K, Yamaji H, Tokshiki M (1987) Microbiol. Biotechnol. 26: 495
12. Nilsson K (1987) Tib-Tech 5: 73
13. Earle WR, Bryant JC, Schilling EL (1953–54) Ann. N.Y. Acad. Sci. 58: 1000
14. Whiteside JF, Whiting BR, Spier RE (1979) Develop. biol. Standard. 42: 113
15. Thornton B, McEntree ID, Griffiths B (1985) Develop. biol. Standard. 60: 475
16. Lydersen BK, Pugh GG, Paris MS, Bhavender PS, Noll LA (1985) Ceramic matrix for large scale animal cell culture. Bio/technol. Jan. 1985, p 63
17. Klement G, Scheirer W, Katinger H (1987) Develop. biol. Standard. 66: 221
18. Knazek RA, Kohler PO, Gullino PM (1974) Exp. Cell. Res. 84: 251
19. Altshuler GL, Dziewulski DM, Sowek JA, Belfort G (1986) Biotech. Bioeng. XXVIII: 646
20. Chresand TJ, Gillies RJ, Dale BE (1988) Biotech. Bioeng. XXXII: 983
21. Schonherr OT, Van Gelder PT, Van Hess PJ, Van Os AM, Roelofs HW (1987) Develop. biol. Standard. 66: 211
22. Katinger H (1988) A tubular biological film reactor concept for the cultivation and treatment of mammalian cells. In: Spier RE, Griffiths B, (eds) Animal cell biotechnology, Academic, London, vol 3, p 240
23. Peters MS, Timmerhaus KD (1981) Plant design and economics for chemical engineers, 3rd edn, McGraw-Hill, p 33
24. Fogler HS (1986) Elements of chemical reaction engineering. Prentice-Hall, p 15
25. Dean AC, Ellwood DC, Evans, CG, Melling J (eds) (1976) Continuous culture 6: Applications and new fields, Elliswood, Chichester, U.K., see e.g. pp. 1–7
26. Kluft C, Van Wezel AL, Van der Velden CA, Emeis JJ, Verheijen JH, Wijngaards G (1983) Large-scale production of Extrinsic (Tissue-Type) Plasminogen Activator from human melanoma cells. In: (ed) Advances in biotechnological processes, Alan Liss, New York, vol 2, p 98
27. Katinger HW, Reiter M, Weigang W, Ernst W, Doblhoff-Dier O, Borth N, Steindl F (1988) A scaleable modular vibrating cage system for enhanced gas/liquid diffusion and cell immobili-
- sation in continuous perfused culture of mammalian cells. Conf. Abs., Engineering Foundation (New York) Conf. on: Cell Culture Engineering, Florida, Jan. 31–Feb., 1988
28. Varecka R, Scheirer W (1987) Develop. biol. Standard. 66: 269
29. Handa A, Emery AN, Spier RE (1987) Develop. biol. Standard. 66: 241
30. Aunins JG, Croughan MS, Wang DIC (1986) Biotech. Bioeng. Symp. No. 17, p 699
31. Croughan MS, Hamel J-F, Wang DIC (1987) Biotech. Bioeng. XXIX: 130
32. Katinger H, Scheirer W (1982) Acta Biotechnologica 2: 3
33. Spier RE, Griffiths B (1984) Develop. biol. Standard. 55: 81
34. Grote J, Susskind R, Vaupel P (1977) Pflugers Arch. 372: 37
35. Warburg O (1923) Methoden. Biochem. Zeitschr. 142: 317
36. Glacken MW, Fleischaker RJ, Sinskey AJ (1983) TibTech 1: 102
37. Whiteside JP, Farmer S, Spier RE (1985) Develop. biol. Standard. 60: 283

Membrane Bioreactors: Present and Prospects

Ho Nam Chang*[1] and Shintaro Furusaki[2]
[1]Department of Chemical Engineering, Korea Advanced Institute of Science and Technology, P.O. Box 131 Cheongryang, Seoul 130–650, Korea
[2]Department of Chemical Engineering, Faculty of Engineering, The University of Tokyo, Bunkyo-ku, Tokyo 113, Japan

* To whom all correspondence should be addressed.

Advances in Biochemical Engineering
Biotechnology, Vol. 44
Managing Editor: A. Fiechter
© Springer-Verlag Berlin Heidelberg 1991

Membrane bioreactors have a very handy in-situ separation capability lacking in other types of bioreactors. Combining various functions of membrane separations and biocatalyst characteristics of enzymes, microbial cells, organelles, animal and plant tissues can generate quite a number of membrane bioreactor systems. The cell retaining property of membranes and selective removal of inhibitory byproducts makes high cell density culture possible and utilizes enzyme catalytic activity better, which leads to high productivity of bioreactors. Enzyme reactions utilizing cofactors and hydrolysis of macromolecules are advantageous in membrane bioreactors. Anaerobic cell culture may be efficiently carried out in membrane cell recycle systems, while aerobic cultures work well in dual hollow fiber reactors. Animal and plant cells have much a better chance of success in membrane reactors because of the protective environment of the reactor and the small oxygen uptake rate of these cells. Industrial use of these reactors are still in its infancy and limited to enzyme and animal tissue culture, but applications will expand as existing problems are resolved.

1 Introduction

In the early 1980s the impact of genetic engineering was largely limited to pharmaceutical proteins such as insulin, interferon and human growth hormone. In the bioprocessing of these high-value-low-volume products more emphasis was placed on downstream processing rather than bioconversion/bioprocesses. However, with the advances of new biotechnology its application is gradually expanding to medium-volume enzyme products and high volume bulk products such as ethanol. In the high volume low value products the productivity of bioreactors plays an important role in determining the economics of bioprocesses.

Traditionally production in the bioprocessing industries has been by batch, which yields lower reactor productivity and higher production costs. The so-called "bioreactor approach" focuses attention on the development of product-specific, high-productivity bioreactors. In order for a bioreactor to have high productivity it is necessary to run the reactor continuously while maintaining high biocatalyst activity. The high cell density does not necessarily lead to high productivity of the reactor because product concentration is very often limited by the product concentration itself. If this is the case, the integration of separation steps with bioconversion or in situ product recovery is advantageous. The most ideal bioreactors will have high reactor productivity with an in-situ separation capability of products.

The bioreactors which fit this description best are membrane bioreactors. Membrane technologies have already contributed a great deal to bioprocessing (Table 1). The use of membranes in bioreactors is still in its infancy, however its potential and its impact will be larger than other applications in bioprocessing. Membrane

Table 1. Use of membranes in biotechnology

1. Air cleaning and sterilization	• HEFA filter
Oxygen enrichment	• silicone membrane
Liquid medium	• ultrafiltration of heat sensitive nutrients
2. Bioreactor	• enzyme, microbial, and animal and plant tissue culture
3. Separation	• reverse osmosis, ultrafiltration, microfiltration,
	• diffusion and electrodialysis
4. Biosensors	• functional membrane

bioreactors are devices in which enzymes, organelles, microbial, animal and plant cells are retained by means of membranes for production of valuable materials, for processes such as wastewater treatment and for analysis used in biosensors.

The biocatalysts can be retained either behind a membrane barrier, in the membrane matrix, or attached to the membrane surface. What distinguishes these bioreactors most from other types is the in-situ separation of products from biocatalysts. The advantages and limitations of membrane bioreactors are summarized in Table 2.

The first membrane bioreactor was used by Gerhardt and Gallup [1] in 1963 for dialysis culture of *Serratia marcescens* of which cell mass increased to 91.9 g l^{-1}. Though Chang [2] pioneered enzyme immobilization by microencapsulation in 1964, the first enzyme membrane reactor was developed by Butterworth et al. [3] in 1970 to carry out starch hydrolysis with α-amylase. For animal tissue culture Knazek et al. [4] cultured human choriocarcinoma cells on mixed bundles of Amicon fibers in 1972. There have been reviews on the specific use of hollow fiber bioreactors for enzyme catalysis [5, 6], microbial cells [7] and animal tissue culture [8, 9]. Two recent reviews on immobilized biocatalysts dealt with membrane bioreactors [10, 11]. Chang [12] briefly reviewed general aspects of membrane bioreactors covering topics from enzymes to animal tissue cultures. This review will extend the scope of the previous review and focus on the engineering aspects of membrane reactor systems in depth and compare them with other types of bioreactors. As a result, the potentials and limitations of membrane reactors for industrial application will be identified.

Table 2. Advantages and disadvantages of membrane bioreactors

Advantages	Disadvantages
In-situ separation of products	Sterilization difficulty
Separation of SRT and HRT	Aeration difficulty
High density cell culture	Scale-up of a single unit is difficult
High productivity	Scale-up merit is small
Phase separation in solvent extraction is easy	Pretreatment of substrate necessary
Small scale economics	

SRT = solids retention time; HRT = hydraulic retention time

2 Membrane Reactor Characterization

2.1 Driving Forces for Separation

Membranes are thin barriers that can perform various degrees of separation using differences in concentration, pressure and electrical potential gradient between the two compartments they separate (Fig. 1). Membrane separations are achieved as a result of combined actions among membrane, solvent and molecules to be separated. In Table 3 are listed various factor affecting separations and unit operations used in the bioprocessing. Bioseparations can be successful only if membrane, solvent and molecules of interest are interacted in harmony. Ultrafiltration, microfiltration, and reverse osmosis are modulated by pressure; dialysis and membrane extractions are carried out using the concentration differences; ionic species or ionized chemicals may be separated by the electrical potential difference using electrodialysis. In many cases membrane separations are based on a single

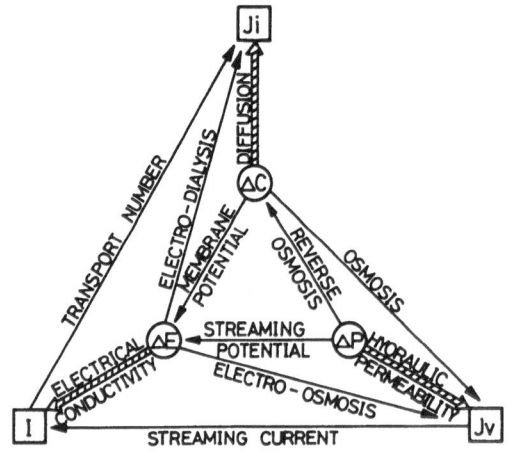

Fig. 1. Relationships among three driving forces in the membrane separation [13]

Table 3. Primary factors affecting bioseparations and unit operations [14]

Factors	Unit operations
Size	Microfiltration, ultrafiltration, gel filtration
Diffusivity	Reverse osmosis, dialysis
Ionic charge	Electrodialysis, ion exchange adsorbent
Volatility	Distillation, vacuum distillation, pervaporation
Solubility	Solvent extraction, precipitation/flocculation
Surface activity	Foam and bubble fractionation
Biological activity	Affinity chromatography
Hydrophobicity	Hydrophobic adsorption

separation mechanism. For instance ultrafiltration uses size as the separation criterion. However, there is a possibility of using two or more separation mechanisms for membrane separations. In this case the separation step will be more complex, but the separation factor will be larger than using a single separation mechanism. In extractive bioconversion/bioprocesses two different fluids can be used: for instance, on one side of the membrane an aqueous solution is used while on the other side an organic solvent or a vacuum can be used. Thus there exist a large number of possible membrane separation processes if we combine primary factors for separation, properties of membrane and molecules to be separated.

2.2 Materials

Membranes are usually made of polymers, but recently ceramics have also been used (Table 4). Polymer membranes are more easily produced than ceramic or stainless steel membranes, but they are structurally weak in comparison with the other two materials. Polymer membranes can be produced very thin and made so as to possess a variety of functions ranging from gas permeation to electrodialysis, but because of structural weakness the membranes need solid support or spacers for use in membrane modules. Ceramic membranes are favored in ultrafiltration processes because they can withstand repeated heat sterilization.

Table 4. Membrane materials [15]

Polymers

(gas separation)
silicone rubbers, hydrocarbon rubbers, polyphenylene oxides, polysulfones, polyesters, acrylonitrile copolymers, polyolefins, siloxane copolymers, aromatic polyamides, polyimides, polycarbonates
(dialysis)
cellulose acetate, regenerated cellulose, polyacrylonitrile, polymethylmethacrylate, polyethylene vinylalcohol, polysulfone, polyamide
(reverse osmosis)
cellulose acetate, modified polysulfones, aromatic polyamides, polyesters, polybenzimidazole, aromatic polyethers
(ultrafiltration)
porous polyethylene, porous polypropylene, cellulose acetate cellulose nitrate, polyelectrolyte complex, polyacrylonitrile, polysulfones, polyamides, polyesters
(ion exchange membranes)
Nafion (sulfonated perfluorocarbon polymer), sulfonated polystyrene, chlorosulfonated polysulfone, polyacrylic acid − zirconium hydroxide

Ceramics

(ultrafiltration) − alumina, alumina silicate
(gas separation) − alumina, alumina silicate

Metals

(ultrafiltration) − stainless steel

Membrane materials to be developed in the future will be those with functional properties which biological membranes possess. If these properties can be incorporated into the membrane material, membranes will find much wider application in bioprocessing.

2.3 Modules

The membrane itself is neither strong enough to build a membrane device alone nor has it enough surface area for mass transfer. Thus it is packaged into a module in flat, spiral, turbular and hollow fiber types. Flat and spiral types need separate spacers to support the membranes. The advantage of these two membrane configurations is that high pressure can be applied. In the case of the spiral type, excessive pressure drop can occur due to the long mass transfer length for one side of the fluid and of the other this can be no problem. Table 5 shows the specific surface areas of various membrane modules. The highest specific surface area is obtained in microcapsules, but packing limitations occur when the capsules are suspended in liquid.

Table 5. Specific surface area of membrane modules

Diameter	Specific[1.] surface area $(cm^2\ cm^{-3})$	Specific surface area $(cm^2\ cm^{-3})$			
		0 = 0.1	0.2	0.3	0.4
Hollow fibers					
50 μm	800	80	160	240	320
100 μm	400	40	80	120	160
200 μm	200	20	40	60	80
300 μm	133	13	27	40	53
Tubular					
1 cm	4	0.4	0.8	1.2	1.6
Flat membrane					
Spiral wound			5–8		
Microsphere					
10 μm	6000	600	1200	1800	2400
20 μm	3000	300	600	900	1200
100 μm	600	60	120	180	240
200 μm	300	30	60	90	120
500 μm	120	12	24	36	48
Taylor vortex					

1. based on a single fiber or microsphere

2.4 Classification of Membrane Bioreactors

Membrane bioreactors can be divided into largely two groups. One category uses the membrane for retaining cells and thus separates the biocatalysts from the solution, resulting in the difference of hydraulic and solid retention times. In this case biocatalysts are suspended in the reactor. The second category immobilizes biocatalysts on the surface, in the matrix or retains them between two membranes. In this case the biocatalysts are not mobile unlike those in suspension. These two groups of membrane bioreactors can be subdivided into several variations depending on whether they are operated on ultrafiltration mode or on diffusion mode; or incorporate a product removal unit or an aeration unit. Figure 2 shows stirred tank type membrane bioreactors where biocatalysts are retained by the membrane but suspended in the fluid. Figure 2a is the simplest membrane bioreactor which was used by Butterworth et al. [3] for hydrolysis of starch using glucoamylase. The problem was that the flux declined very sharply with time due to the concentration polarization on the membrane surface and the enzyme leaked through the membrane pores. Currently, concentration polarization can be minimized by employing tangential flow filtration. Thus nearly all membrane devices use tangential flow. Some of the outlet stream can be bled off without going through the membrane for continuous processing (Fig. 2b). The biocatalysts can be placed separately from the reactant and are recycled to reduce the film resistance (Fig. 2c). Products can be removed either by dialysis or by extraction with solvent or by pervaporation (Figs. 2d and 2e). Combining the reactor 2b and 2e can produce reactor 2f. Membranes can also be used for aeration purposes (Fig. 2g).

The immobilized biocatalyst in, on or between the membranes can form several types of reactor configurations. Figure 3a represents the simplest type, but this

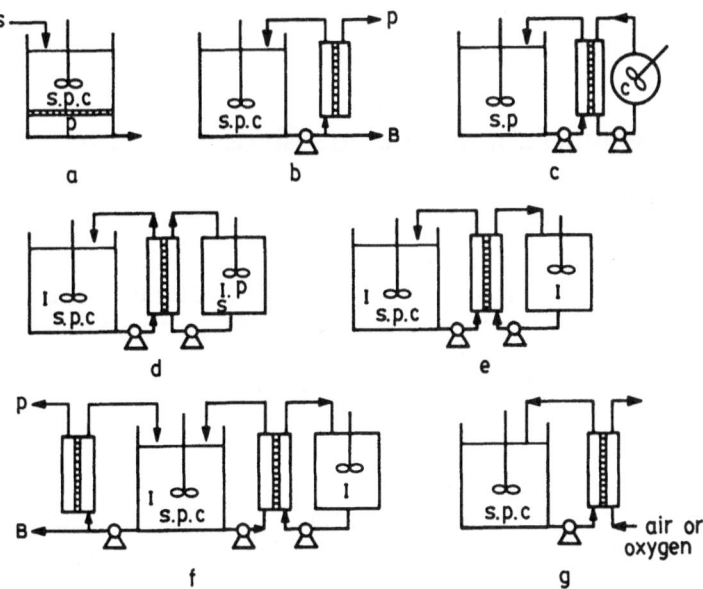

Fig. 2 a–g. Membrane bioreactors for suspended biocatalysts: **(a)** ultrafiltration cell (U) **(b)** membrane recycle with bleed (U) **(c)** membrane recycle (D) **(d)** dialysis (D) **(e)** extractive operation (D) **(f)** ultrafiltration with extraction or dialysis (U + D) **(g)** aeration (D) U: ultrafiltration mode S: substrate P: product C: biocatalysts D: diffusion mode I: inhibitor

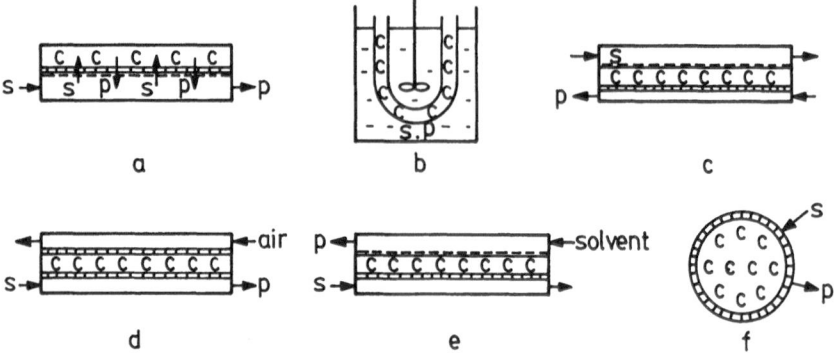

Fig. 3 a–f. Membrane bioreactors for immobilized biocatalysts: **(a)** hollow fiber (D) **(b)** hollow fiber beaker (D) **(c)** hollow fiber (U) **(d)** dual hollow fiber (D) **(e)** hollow fiber with extraction (D) **(f)** microcapsules (D)

reactor only receives oxygen through the incoming fluid. Thus when the liquid has a limited solubility for oxygen, reactions requiring oxygen cannot be carried out. To solve this problem the dual hollow fiber bioreactor was invented employing an oxygen-permeating membrane on one side and an ultrafiltration membrane on the other side (Fig. 3d). Figure 3e shows that products can be stripped by solvents to reduce product inhibition. Microencapsulation is one way of immobilizing biocatalysts, but it requires a reactor such as stirred tank reactor where these capsules can be suspended (Fig. 3f).

3 Transport Phenomena in Membrane Bioreactors

3.1 Fluid Mechanics

Fluid mechanics plays an important role in membrane devices. In mass transfer devices virtually all mass transfer is either governed by or related to fluid mechanics [16]. Some important problems in membrane systems related to fluid mechanics are:
1. flow distribution problems in membrane devices.
2. pressure drop in a hollow fiber with pulsating flow and with permeation.
3. starling effect.
4. fluid mechanics in membrane spacers.
5. flux decline with concentration polarization.
The simplest but most important problem in membrane devices is how to distribute incoming fluid evenly into the multi-channels or hollow fibers of membrane devices. If the flow is not distributed evenly in flat or spirally wound membrane modules, channeling will take place like in packed bed systems. In other words, some parts of the membrane surfaces will not be utilized for mass transfer. For this reason flat membrane devices use preformed spacers that form grooves for flat membranes. Then even flat membrane devices work like hollow fiber systems (Fig. 4a). The

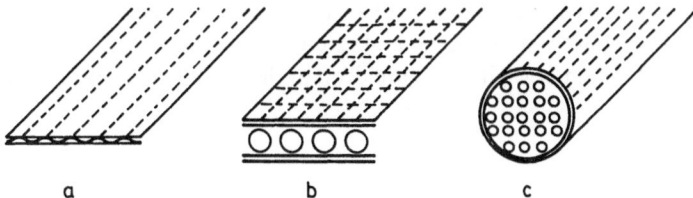

Fig. 4 a–c. Flow distributors in membrane modules: (**a**) multigrooved channels (**b**) turbulence promoters (**c**) hollow fibers

second alternative in flat membrane systems is the use of spacers. The primary role of spacers is to keep membranes apart, but the most important roles are to make fluid distribution even for maximal utilization of the mass transfer area and to promote mixing of fluids between two membranes to result in improvement of mass transfer efficiency by reducing mass transfer boundary layer thickness [17]. In the case of hollow fibers, fluid distribution problem was dealt with by Chang and Park [18]. The governing fluid mechanics equation for hollow fiber inlet and outlet head spaces were solved numerically and the fluid distribution profile was obtained. Figure 5 shows that the distribution is a function of the Reynolds number, pressure drop in the fiber, head height and diameter of the module.

The pressure drop in the hollow fibers normally does not present a problem in membrane bioreactors unless the fiber is very long. The longest hollow fiber used

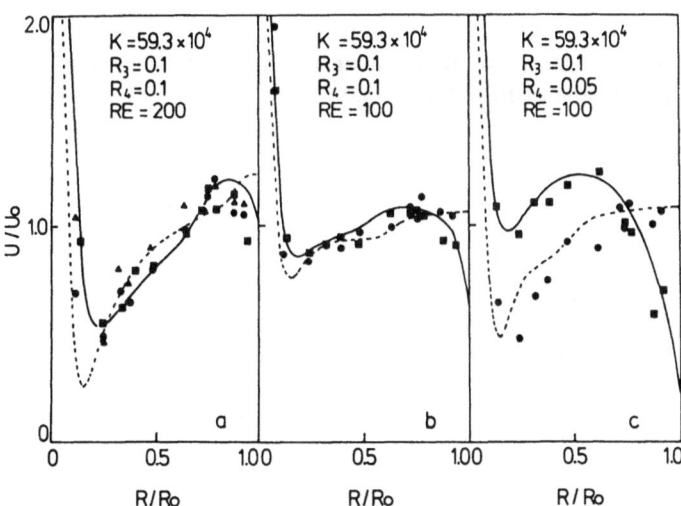

Fig. 5. Flow distribution in a hollow fiber module [18]. R/R_0 represents dimensionless radial distance from the center. K: pressure factor, R3: diameter ratio of hollow fiber to the module. R4: ratio of head clearance to the module diameter. *Dotted* and *solid lines* show the results with cylindrical and conical head shapes

in the module is 63.8 cm. If the diameter of the hollow fiber is 50 μm, viscosity 1 cp and the average velocity in the fiber is 42.5 cm s^{-1}, then total maximum pressure drop will be 245 mm Hg. The pressure drop in the module of multi-fibers is given by

$$\Delta P = \frac{8\mu L}{\pi R^4}\left(\frac{Q}{n}\right),$$ (1)

where Q is flow rate (cm^3 s^{-1}) and n is the number of fibers. In the case of spirally wound modules the fluid path in the membrane is long enough to cause excessive pressure drop especially when the gap between the membranes is small. The pressure drop along the fibers in the presence of permeation is not the same as that without permeation. In the presence of permeation the pressure drop is smaller than in its absence because the amount of fluid decreases with the length of the fiber. This is expressed as [19]

$$\Delta P = a[1 - \exp(-\alpha L)] + \beta L .$$ (2)

The first term represents the pressure drop due to ultrafiltration. The Starling effect occurs because of the pressure differentials between the inlet and the outlet of the fibers (Fig. 6). If the resistance through the membrane pores is small, then the amount of fluid transversing the membrane will be large. Waterland et al. [20] reported the localization of enzymes on the downstream side of the hollow fibers due to the Starling effect. This effect is well known in capillary blood flow, but in the case of hollow fibers the phenomenon has only been explained conceptually and not yet in quantitative terms. In a membrane system where a flat membrane is employed, concentration polarization on the membrane surface is a common phenomenon. Membrane spacers promote mixing of fluids between the two membranes [21–23]. The role of membrane spacers and its mass transfer effect are reviewed elsewhere [17].

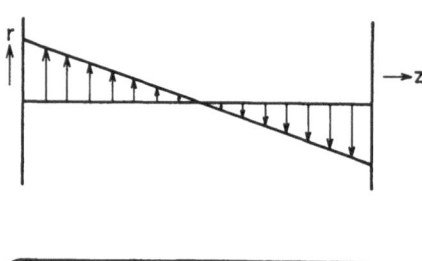

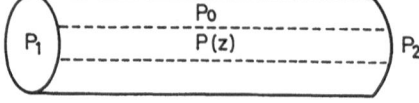

$\Delta P(z) = P(z) - Po \; (= \dfrac{P_1 + P_2}{2})$

Fig. 6. Starling effect in hollow fibers

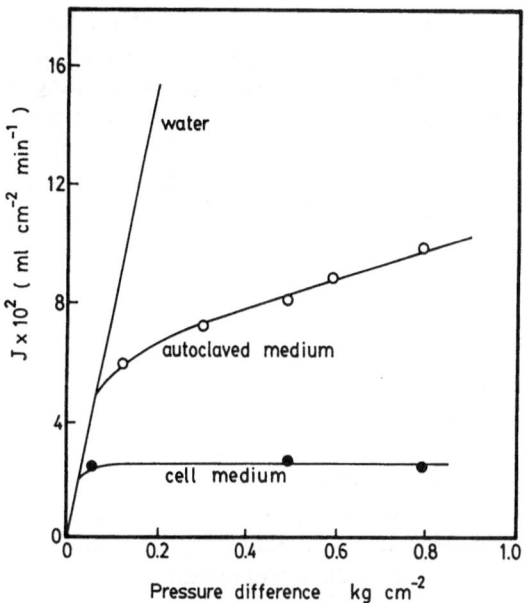

Fig. 7. Comparison of fluxes of pure water, reaction medium and cell containing fluids

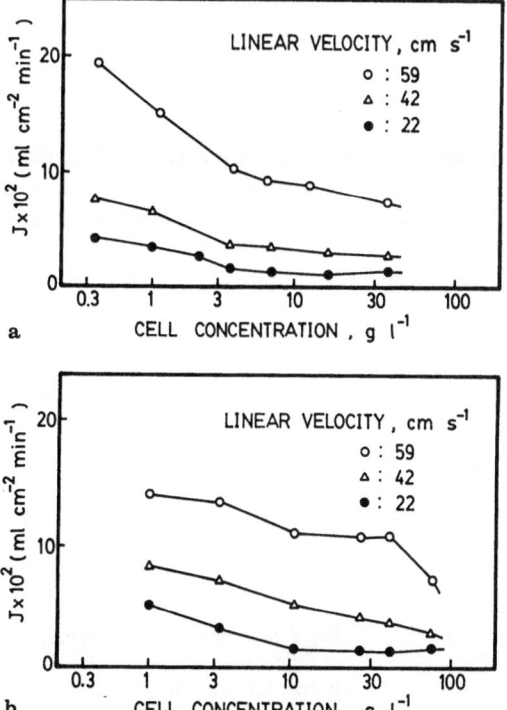

Fig. 8 a, b. Fluxes of cell containing fluids depending on cell concentrations [25]: **(a)** *E. coli* **(b)** *S. cerevisiae*

The use of larger pore membranes allows the flow passage of 1–10 Å molecules and retention of proteins, other macromolecules and cells. As the filtration proceeds, a layer of solute-rich fluid builds up at the interface resulting in concentration polarization. When enzymes or cells account for the concentration polarization layer, the osmotic pressure is typically negligible. But the accumulating layer can create mass-flow resistance. Figure 7 shows flux declines of pure water, process fluids and the cell containing fluids [24]. The flux depends on cell type and concentration, and the flow rate in the membrane device (Fig. 8) [25].

3.2 Mass Transfer

The advantage of the in-situ separation capability that membranes posses is accompanied by the possibility of diffusional resistance. This, of course, can be a problem in other types of reactors containing immobilized biocatalysts. This mass transfer resistance is caused by the membrane itself and film boundary layers on both sides of the membrane. The latter are caused by poor mixing of fluids near the membrane surface and are unavoidable unless carefully designed modules are used. Membrane resistance can be overcome by the use of ultrafiltration through the membrane, but decreasing flux with time makes long term operation infeasible. Between the diffusion and ultrafiltration operations periodic mixing of fluids between the two compartments can significantly reduce mass transfer resistance imposed by the membrane (Fig. 9). Figure 10 shows various modes of periodic exchange by convection between two compartments. This can be done by pressure pulsing [26] or by ultrafiltration pulsing [27, 28]. Figure 11 shows that the conversion of lactose can be significantly improved [29]. This technique can eliminate most membrane resistance when it becomes serious, and allows the long-term operation of enzyme reactors.

For aerobic reactions requiring oxygen, the oxygen transfer rate in hollow fiber reactors is not sufficient to carry out reactions requiring oxygen such as the enzymatic oxidation of glucose or aerobic culture of microbial cells and animal and plant tissues. The introduction of dual hollow fiber bioreactors has solved much of the oxygen limitation problem in membrane bioreactors [30]. Another

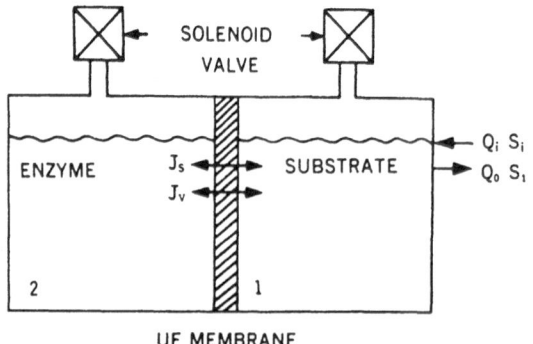

Fig. 9. Schematic diagram of periodic fluid exchange across the membrane for membrane enzyme reactors

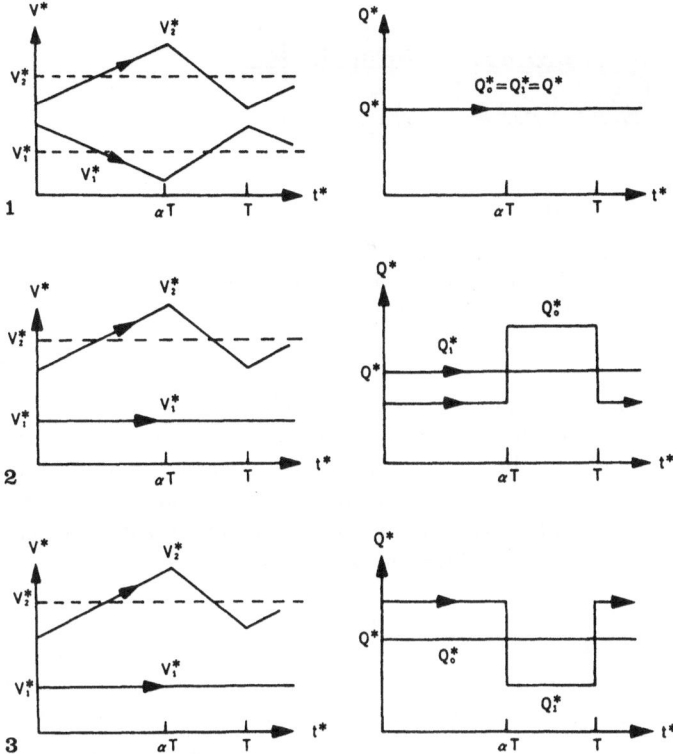

Fig. 10. Various methods of fluid exchange between enzyme and substrate compartments: **(1)** pressure swing **(2)** ultrafiltration swing at outlet **(3)** ultrafiltration at inlet. $V1^*$ and $V2^*$ represents the amount of fluids in each compartment with time

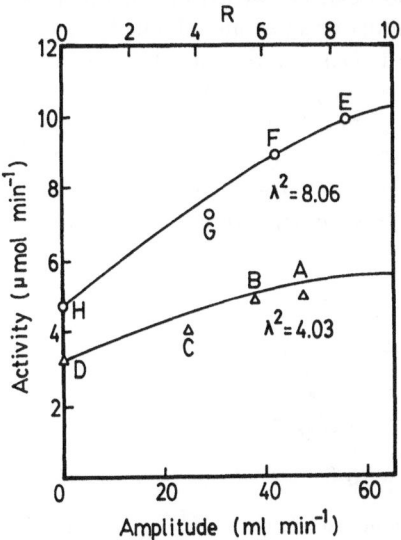

Fig. 11. Effect of pulsing on the apparent enzyme activity for enzyme reactors. Zero amplitude is the control value. For further details see Ref. [28]

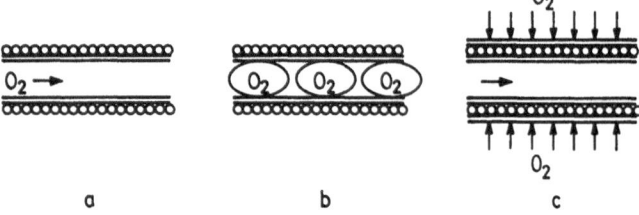

Fig. 12 a–c. Method of oxygen supply in hollow fiber bioreactors: **(a)** O_2 dissolved in homogeneous fluids **(b)** O_2 loaded fluids and water **(c)** O_2 directly supplied to immobilized cells through membrane without going through liquids (dual hollow fiber bioreactors)

method of oxygen supply is to use oxygen carrying bubbles as oxygen carriers in the hollow fibers [31]. Thus slug flow is utilized. The efficiency of this method lies between the diffusion and dual hollow fiber methods. Figure 12 depicts three modes of oxygen supply in hollow fiber reactors. In the case of animal cell culture direct aeration cannot be used because of bubble formation. In this case a porous polypropylene tube or a silicone membrane is used.

3.3 Reactor Operation

Membrane reactors which use biocatalysts in suspension are usually operated in the continuous stirred tank mode and this operation is favorable when reactions are inhibited by substrate or product. This configuration is good when reactions requiring oxygen are carried out since it is not easy to supply sufficient oxygen in membrane bioreactors operating on a plug flow mode. If the reaction is of the Michaelis-Menten type, the reactor can be operated in a plug flow mode. Many enzyme reactions of simple kinetic expression can be carried out in this mode (Fig. 13). Reactions requiring oxygen or product or substrate inhibited reactions can be carried out in a rapid recycle mode (STR). Another advantage of rapid recirculation lies in the reduction of film resistance present in the boundary layer near the membrane surface. It is worth noting that lengthening residence time in the membrane reactor where the biocatalysts are immobilized is not always beneficial even when operated in a plug flow mode because of large film resistance. With method 12a the maximum dissolved oxygen in equilibration at one atmos-

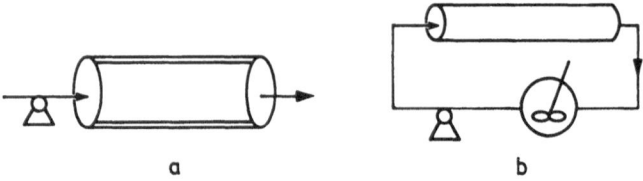

Fig. 13 a, b. Method of reactor operation: **(a)** Plug flow mode **(b)** Rapid recycle mode

pheric pressure is only 0.25 mM. If the oxygen demand rate is Qo_2, then the supplied oxygen will be consumed within a residence time θ

$$\theta = \frac{0.25}{Qo_2}. \tag{3}$$

If Qo_2 is 100 Mm $l^{-1} h^{-1} O_2$, it will be depleted if the residence time is longer than 9 seconds. However, method 12c can supply 205 Mm $l^{-1} h^{-1} O_2$ when the inner tubing diameter is 0.0737 cm and also it has been shown that the silicone tube itself is not resistant to O_2 permeation [32, 33].

4 Enzyme Reactors

4.1 Application Systems

Membrane bioreactors are considered ideal for systems where the enzyme and any cofactors must be contained and the product is removed. Enzymes can carry out oxidation-reduction, transfer of functional groups, hydrolysis, addition to double bonds, isomerization reactions and formation of bonds with ATP cleavage. Substrates for enzyme reaction can be polymers of large molecular weight such as starch or cellulose, small water-soluble or oil-soluble molecules. Enzyme kinetics can exhibit simple Michaelis-Menten, reversible Michaelis-Menten, competitive or noncompetitive inhibition kinetics. Or enzymes can require one or two substrates for the reaction. These factors governing enzyme reactions are summarized in Table 6.

Table 6. Various factors for enzyme reactions for membrane bioreactors

Reaction type	Oxidation-reduction, transfer of functional groups hydrolysis, addition of double bonds.
	Isomerization reactions and formation of bonds with cleavage.
Substrates	Polymers (starch, protein, cellulose), water-soluble or oil-soluble.
	Cofactors (metal ions and NAD, FAD. ATP etc).
Kinetics	Michaelis-Menten, reversible MM, competitive or noncompetitive inhibition kinetics.
Stability	Several hours to several months.
Operating conditions	Low-high pH, low-high temperature.

4.2 Reactor System Development

Enzyme reactor systems can be developed to meet the situations listed in Table 6. May be simple bioconversion of a small molecule can be carried out in a membrane reactor without difficulty, but water-insoluble substrates cannot be converted in a simple membrane reactor. Thus we can categorize the types of reaction which a membrane reactor is able to carry out.

4.2.1 Hydrolysis Reaction of High Molecular Weight Substrates

Hyxdrolysis reactions of starch [3, 34, 35], cellulose [36, 37] and protein [38–40] can be carried out efficiently in membrane bioreactors. For this type of reaction, membrane reactors have an advantage over other types of reactors. If enzymes are not intended to be recovered, simple batch reaction will do. But if enzymes are to be reused, maybe a membrane reactor is the only candidate.

In 1969, Butterworth et al. [3] carried out starch hydrolysis using α-amylase in a stirred ultrafiltration cell. It was observed that the flux declined with time due to concentration polarization. Enzyme activity also decreased, possibly due to enzyme leakage. The problem of the ultrafiltration cell depicted in Fig. 2a is the application of dead-end filtration. Nowadays dead-end filtration is rarely used; instead, tangential (or crossflow) filtration is used. This new filtration method greatly improved the situation. Protein and cellulose reactions were also carried out. The most appropriate type reactor will be that shown in Fig. 2b. Centrifugal filtration [41] developed recently is even more efficient than tagential filtration, but it is difficult to use in large scale systems.

4.2.2 Simple Bioconversion Reactions

Simple bioconversion reactions can be carried out in a batch mode or in a continuous mode. If we consider the effectiveness, the reaction should be carried out in plug flow reactor where biocatalysts are suspended or immobilized. Waterland et al. [20] adsorbed β-galactosidase on the sponge region of a hollow fiber module to carry out hydrolysis of ONPG. Horvath and Solomon [42] immobilized trypsin on the inner surface of a tubular membrane. Dynamically formed enzyme gel membranes can also be used [43]. Kawakami et al. [44] compared the performance of a beaker-type hollow fiber reactor where in one case urease was immobilized in the reservoir and substrate on the shell side and vice versa and found that the former was better.

4.2.3 Reactions Requiring a Cofactor

Reactions requiring a cofactor are possible in a reactor where two substrates including the cofactor are suspended. Because of the high price of cofactors it is necessary to recover cofactors after the reaction. Thus when we use membrane reactors, we attach long-chain polymers (usually PEG) to a cofactor so that it can be retained in the reactor (Fig. 14) [45, 46]. The chain length of polymers can affect the coenzyme activity and retention of enzymes in the membrane reactor as well as the viscosity of the medium, which ultimately limits the amount of ATP in it. Ishikawa et al. [47] investigated the production of glucose-6-phosphate by regenerating ATP in an ultrafiltration hollow fiber reactor. NAD regeneration was studied by Miyawaki et al. [48] using alcohol dehydrogenase and lactate dehydrogenase immobilized in hollow fibers.

coenzyme dependent reaction

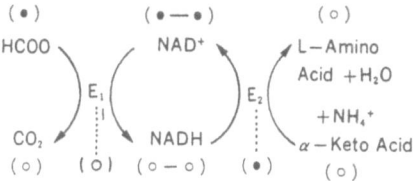

enzyme membrane reactor

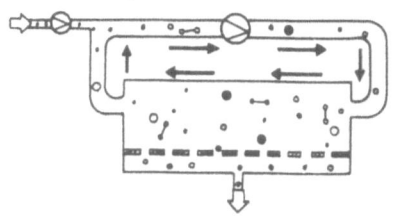

Fig. 14. Enzyme membrane reactor concept for continuous enzymatic synthesis with co-enzyme regeneration [45]: E2 = L-leucine dehydrogenase, E1 = formate dehydro-genase, MAD, NADH-PEG-10,000 NADH, L-Leucine, α-ketoisocaproate

4.2.4 Reactions Requiring Oxygen

Oxygen can be transferred to bioreactors by direct bubbling of air or oxygen, surface aeration or through membranes. Because of the low solubility of oxygen in water, it is not efficient to immobilize an enzyme in a support. The effectiveness will be much lower because oxygen will be rapidly consumed near the surface of the support [49]. If we look at the oxygen penetration depth, this becomes clear. This means that the reaction should be carried out in batch reactor where substrate and enzyme are suspended with an oxygen supply. But in this case the enzyme deactivation may be accelerated due to the direct contact of air with the enzyme protein because protein can be denatured by exposure to air. After the reaction is completed, the product can be recovered.

The second alternative is that the reaction can be carried out in a plug flow reactor where oxygen is supplied in parallel. An example of this type reactor is the rotating biological contactor. Chang et al. [32] performed the glucose oxidation reaction using a dual hollow fiber reactor. The efficiency of the reactions was about 13 times better than with a rotating packed disk reactor [50].

4.2.5 Water-Insoluble or Sparingly Soluble Substrates

Many enzyme substrates are either insoluble or sparingly soluble in water. Since enzymes usually function in an aqueous environment, it is necessary to devise a reactor that enables good contact to be made between these substrates in water. A good example of this is the hydrolysis of fat to form water-soluble glycerides and sparingly soluble fatty acids. Gylcerides are removed from the reaction mixture with water (Fig. 15) [51, 52]. Many enzyme reactions are inhibited either by substrates or by products. If this is the case, we can use an organic solvent or an aqueous second phase on one side of the membrane to perform extractive

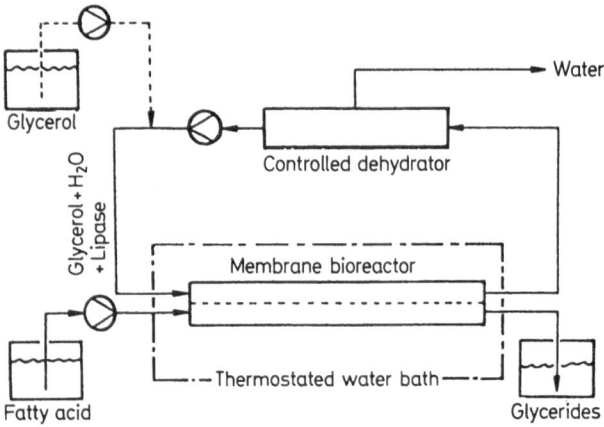

Fig. 15. Synthesis of glycerides in membrane bioreactors [51]

bioconversion and on the other side we use enzymes in water [53]. Separation of α- and β-naphthyl sulfates was investigated with the enzyme hydrolysis of the β-ester. Thus the mixture of α-ester and α-naphtol could be changed to the mixture of α-ester and β-naphthol. A liquid membrane of decanol supported by porous polytetrafluoroethylene was used to separate the β-naphthol [54].

4.3 Mass Transfer and Reactor Operation

When we use reactor systems where biocatalysts are suspended and ultrafiltration is the reactor operation mode, CSTR and PFR type reactors are not advantageous because substrates can leave the reactor unconverted. In this case the conversion should be carried out in the batch mode and only after the reaction is complete should filtration or recovery of the enzymes start [55]. Figure 16 shows the production of aspartic acid using *E. coli* aspartase. For a continuous mode

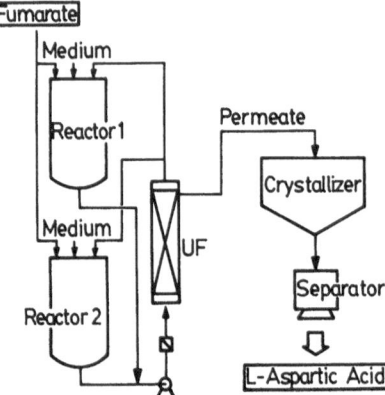

Fig. 16. Production of L-aspartic acid using a membrane bioreactor [55]

operation, a tubular type reactor should be used and at the end of the tubular reactor the reaction will be nearly completed and filtration can begin. For a first order reaction it is better to use plug flow type reactor than CSTR. In order for a reactor to be of plug-flow type, we have to increase the reactor length or use longer residence times. In the latter case, the reaction may be inefficient because of the thick mass transfer boundary layer. Thus the overall reaction rate may be improved by rapidly recirculating the substrate solution through the reactor (Fig. 13b).

4.4 Enzymes Immobilized on the Matrix of Membranes

Biocatalysts can be fixed on membranes. Invertase was immobilized on a polyethylene vinyl alcohol copolymer membrane and amino acetalization was carried out [56, 57]. High enzyme activity was achieved. Similar immobilization of glucoamylase was not successful since the activity decreased significantly with immobilization. The same authors studied immobilization using a polyvinyl alcohol membrane crosslinked by UV irradiation [58]. Furusaki et al. [59] tried electrophoretic immobilization of urease and aspartase on porous membranes and carried out simultaneous reaction and migration of substrates and/or products. Lee and Hong [60] showed the effect of electrophoretic separation with a membrane reactor. Alkali phosphatase was immobilized asymmetrically in a polyacrylamide membrane by applying an electric field during the polymerization process [61]. Ichijo et al. [62] proposed a filter-like knitted membrane made of polyvinyl alcohol superfine filaments to adsorb enzymes or fungi. Gianfreda et al. [63] stabilized enzymes in the ultrafiltration membrane enzyme reactors by adding poly-electrolytes.

Recently Chang and his associates [64, 65] have developed a method for high density enzyme immobilization by growing cells between two hollow fibers until they reach their maximum density and then using them as enzyme reactors. This method has been applied to glucose isomerization and acrylamide production from acrylonitrile.

4.5 Chronological Development of Enzyme Reactors

Membrane enzyme reactors emerged in the late 1960s when commercial membranes were first developed and enzyme immobilization technology became useful for commercial practice. The first enzyme reactor studied by Butterworth et al. [3] was used for the hydrolysis of starch using α-amylase for which there are no other immobilization methods. One of the recent triumphs in membrane enzyme reactors is the production of amino acids with simultaneous regeneration of NADH which has been commercialized by the Degussa Company in Germany [66]. This membrane enzyme reactor has advantages over other immobilization methods. In Japan the Kao Corporation [67] investigated the so-called sandwich reactor for hydrolyzing triglycerides. Nitto Electric Industries [68] immobilized cyclomaltodextrin glucanotransferase (CGTace) on a hollow fiber membrane and investigated production of cyclodextrin from starch. Denki Kagaku Industries,

Mitsui Petrochemical Industries and Institute of Microbiology [69] investigated cooperatively the production of sorbitol and glucanate from glucose by using a negatively charged membrane. Ajinomoto [55] investigated production of L-aspartate with a reactor accompanied by a UF module. Production of L-aspartate by membrane reactors is said to be operated also by several companies, e.g. Mitsubishi Petrochemical Co. and Showa Denko Co.

5 Microbial Cell Cultures

A membrane reactor was first used by Gerhardt and Gallup [1] in 1963 for the dialysis culture of *Serratia marcescens*. The final cell mass reached $8.3 \, \text{g} \, \text{l}^{-1}$ without dialysis, but when the dialysis volume was twice the size of the reactor, the cell density increased to $18.2 \, \text{g} \, \text{l}^{-1}$ and when the ratio was $10:1$, the cell mass increased to $91.9 \, \text{g} \, \text{l}^{-1}$. This example clearly shows the effectiveness of the membrane reactor in cell culture. Microbial cells can be immobilized in membrane matrices, cells can be retained in membrane ultrafiltration units (membrane recycle reactors), and inhibitory substances can be removed through the membrane with running water (dialysis) or solvent (extractive bioprocessing).

5.1 Dialysis and Extractive Bioprocessing

When cells are grown in batch culture, the biomass and product reach a stationary phase after a certain period and do not increase further. This happens for two reasons: substrate depletion and inhibitor accumulation. When the first reason is the problem repeated addition of substrate will increase the biomass and product unless the substrate itself inhibits growth and production. However, when inhibitor accumulation is the cause, the substrate is not completely utilized and thus the addition of substrate will not solve the problem; instead, the removal of products through dialysis or ion exchange becomes necessary [70]. Gerhardt and his associates [71] carried out extensive research on dialysis culture. In order for dialysis culture to be effective, the dialysis tank should be much larger than the main containment; furthermore, the use of membranes in real bioprocess systems is rather difficult. As a result, the dialysis cultures have been limited to laboratory scale. Recently an extractive bioprocessing of ethanol with tri-*n*-butylphosphate (TBP) has been performed [72]. A $500 \, \text{g} \, \text{l}^{-1}$ feed glucose medium was successfully processed by immobilized yeast cells to yield a productivity of ethanol of $48 \, \text{g} \, \text{l}_{\text{gel}}^{-1} \, \text{h}^{-1}$. For practical purposes regeneration of the solvent is necessary. The ethanol process was studied with continuous separation by pervaporation [73–76]. Microporous polypropylene and polytetrafluorotetraethylene were used as the pervaporation membrane and a 6–8 times more concentrated ethanol solution was obtained as compared to the medium, i.e. the permeated solution contained about 25% ethanol.

5.2 Matrix Immobilization

Cells can be retained in the shell side of hollow fiber reactors. They proliferate in the interstitial space between the fibers and in the spongy region of the hollow fibers. Nutrients are supplied through the membrane from the lumen side, and products are removed in the opposite direction.

5.2.1 Anaerobic Cultures

Vick Roy et al. [77] immobilized *Lactobacillus delbreuckii* in the shell side of a hollow fiber module for lactic acid production. Final cell densities in the shell side were apparently as high as 480 g l^{-1} DW (dry weight). The volumetric productivity of lactic acid was 20 times that of batch process modes, but fibers were damaged by the growing cells so that the cells were able to penetrate the walls and contaminate the recycle vessel. Thus long-term operation was not feasible. Inloes et al. [78] and Mehaia and Cherylan [79, 80] carried out ethanol process using *Saccharomyces cerevisiae*, but continuous operation was not successful due to the accumulation of CO_2. Park and Kim [81] ultrafiltered the substrate solution through the membrane from the lumen side of hollow fibers. As a result, washout did not occur until the dilution rate became twice the growth rate. Ooshima et al. [82] used an immobilized yeast silicone rubber composite membrane for ethanol production. Methane was produced from *Methanobacterium thermoautotrophicum* immobilized on the surface of hollow fibers using CO_2 and H_2 as the substrates [83].

5.2.2 Aerobic Cultures

Mattiasson and Ramstorp [84] immobilized *Pseudomonas denitrificans* in an artificial kidney module for continuous water nitrate reduction. In the effluent of the hollow fiber reactor no cells of *P. denitrificans* were detected during the first 48 h of operation, after which the leakage increased rapidly. Inloes et al. [85] immobilized recombinant *E. coli* on a single hollow fiber. The cells propagated to extremely high densities, typically more than 10^{12} cells per ml of accessible void volume. The productivity of β-lactamase in the reactor was 100 times more productive than in a shaker-flask culture on a reactor-volume basis. The cells grew well in the spongy region of the fiber wall, but 6 h after inoculation, the cells were detected in the reactor effluent, which meant that the leaked cells were growing on the lumen surfaces of the fibers. Sometimes nutrient flow stopped because of occlusion of the lumen cross section. Vinegar was produced by *Acetobacter rancens* cells fixed on the surface of polypropylene hollow fibers [86]. Ethanol medium was transported along the outside of the fiber while oxygen was supplied through the inside of the fiber. The productivity was 0.2 g l^{-1} h^{-1} of acetic acid in the hollow fiber reactor compared with 0.02–0.08 g l^{-1} h^{-1} observed in a traditional surface culture for vinegar production. Even if the cell leakage problem were solved, these reactor types are not suitable for aerobic processes because of inadequate oxygen supply.

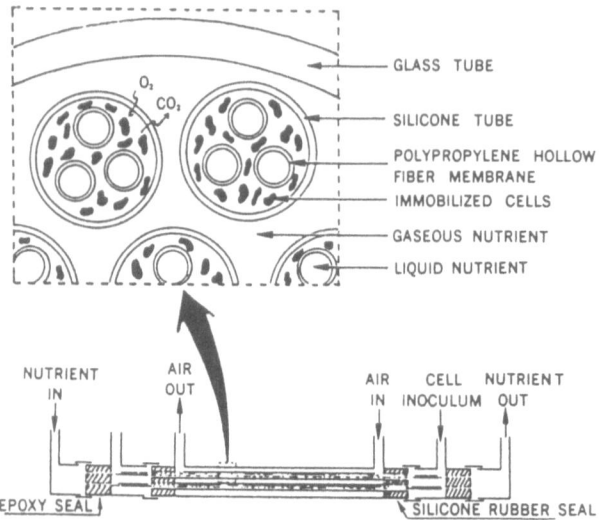

Fig. 17. Schematic diagram of a dual hollow fiber bioreactor

5.2.3 Dual Hollow Fiber Bioreactors (DHFBR)

For supply of large amount of oxygen in ordinary hollow fiber bioreactors it is necessary to use the scheme shown in Fig. 2b or to rapidly recirculate the solution (Fig. 13b). But these two methods are difficult in practice. Robertson and Kim [87] designed a dual hollow bioreactor system (DHFBR) for aerobic cell culture. The reactor consisted of multi polypropylene fibers in which three inner silicone hollow fibers for oxygen transport were placed. *Streptomyces aureofaciens* was grown to produce tetracycline, but the production declined sharply after 3 d when the growth medium was switched to the production medium. Figure 17 shows a

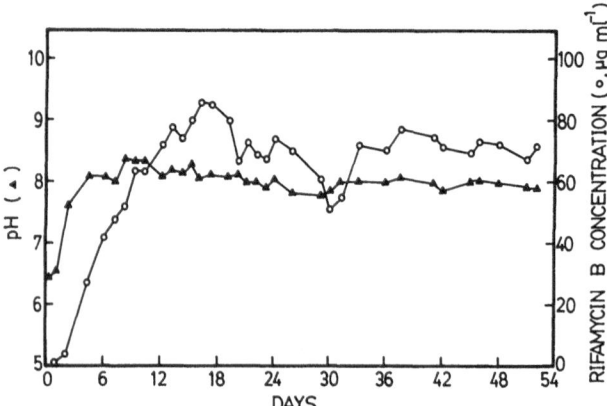

Fig. 18. Continuous production of rifamycin B in the dual hollow fiber for rifamycin B production [88]

schematic diagram of dual hollow fiber bioreactor used by Chang and his associates [88]. Using a slightly modified system Chang et al. [88] produced rifamycin B for more than 50 d using *Nocardia mediterranei* (Fig. 18). This means that long term production of antibiotics or citric acid was possible in DHFBR [89, 96]. The characteristics of this reactor is that minimal aeration is needed in terms of cost as compared to a batch process; minimal sterilization is needed since air and the substrate solution does not come into contact with the cells directly; and furthermore the volumetric productivity is very high (10–30 fold) in comparison with that in a batch culture. If this reactor system is successfully scaled-up, many aerobic products currently produced by batch processes will be economically produced using this reactor. While rifamycin B concentration in DHFBR was only 10% of the batch system, Chung and Chang [89] showed that citric acid can be produced in DHFBR with a yield of 90%, and a 30-fold batch productivity compared with a yield of 40% in the shake flask system (Table 7). Chang and his associates immobilized a variety of microbial cells in the DHFBR and consistently

Table 7. Citric acid productivities and concentration by various methods [89]

Method	Reactor residence time (d)	Initial sucrose conc. (g l^{-1})	Citric acid conc. (g l^{-1})	Yield (%)[a]	Vol. Prod. (g l^{-1} h^{-1})	Reference
1. Continuous culture	0.7 (7.5)	50 (135)	6.5 (20.5)	45	0.4 0.11	[90]
2. Disc reactor (medium replacement)	15	140	80	57	0.22	[91]
3. Air-lift reactor (continuous with Ca-alginate immo.)	(12)	150	12 (320)	26 (60)	0.77[b] (0.2)	[92]
4. Air-lift reactor (batch with Ca-alginate immo.)	24	150	42.7	50	0.074	[93]
5. Tower reactor with two-stage (continuous with polyacrylamide immo.)	6.3	100	24		0.16	[94]
6. Repeated batch (shake flask 2nd run)	7 (14)	140 (140)	77 (63)	70 (61)	0.46 0.19	[95]
7. DHFBR (with pure oxygen and ND medium)[c]	0.84 (12)	60 (60)	26 (18)	80–90 (40)	1.3 0.06	[89]

The value in the parentheses represent the results of comparable batch process.
[a] Yield (%) = produced citric acid per consumed sugar: g g^{-1} × 100
[b] specific unit ref to gel weight: g g^{-1} h^{-1}
[c] ND = nitrogen deficient

Fig. 19. Electron micrograph of *Nocardia mediterranei* grown in dual hollow fiber bioreactor for rifamycin B production. The dry cell weight reached 500 g l⁻¹ [88]

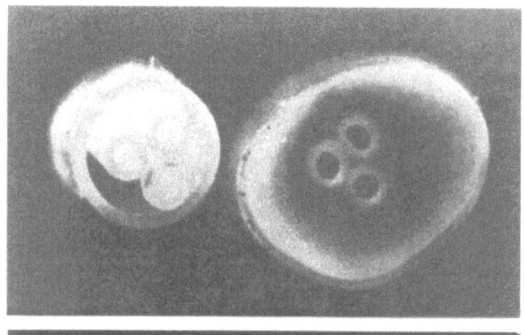

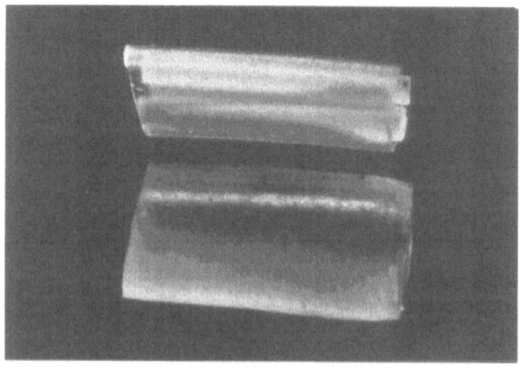

Fig. 20. Expansion of silicone tube by *Aspergillus niger* immobilized in the dual hollow fiber bioreactor for citric acid production [89]

obtained high productivity [64, 65, 89, 96–99]. The higher productivity of DHFBR compared to other types of reactors is attributed to the high cell density achieved in the system. Figure 19 shows rifamycin B producing *N. mediterranei* cells that were tightly packed. In the production of rifamycin B and in the culture of *E. coli* 500–550 g dry weight of cells were obtained. This is the highest cell density ever achieved in all immobilized cell systems. This high density comes from better oxygen supply than in a liquid suspension system since the solubility of oxygen is very poor in liquids. The empty space between the two membranes is a better place than the polymer support that has pores smaller than microbial cells. The difficulties associated with DHFBR are that cells can break hollow fibers *(E. coli)* and fungal growth can expand the fibers to restrict flow path (Fig. 20). Thus ceramic fibers which have enough strength can avoid this difficulty, but because of the thickness, mass transfer can become a problem.

5.3 Membrane Recycle Reactors

Cell recycling has been used in wastewater treatment systems for a long time as a means of increasing biomass. Theoretical analyses on the continuous operation with biomass feedback were performed by Herbert [100] and Pirt [101]. Currently gravity settling practiced in waste water treatment, centrifugation and membrane filtration are used to recycle cell mass.

5.3.1 Anaerobic Cultures

Membrane recycle reactor systems have been widely investigated for ethanol production. Margaritis and Wilke [102] used a rotor reactor equipped with a porous membrane to separate yeast cells from the medium. The productivity of ethanol was $23–28 \text{ g l}^{-1} \text{ h}^{-1}$. Rogers et al. [103] used microchannel filter to retain *Zymomonas mobilis* for rapid ethanol production. The ethanol productivity increased to $120 \text{ g l}^{-1} \text{ h}^{-1}$. Nishizawa et al. [104] carried out ethanol production using dry baker's yeast in the reactor equipped with a cross-flow filter (hollow fiber module). The ethanol productivity was $27 \text{ g l}^{-1} \text{ h}^{-1}$. A microporous hollow fiber module was used to separate yeast cells to keep the cell concentration high and recycle the cells to the reactor [105, 106]. Cherylan and Mehaia [107] used the membrane cell recycle reactor system to obtain productivity of $120 \text{ g l}^{-1} \text{ h}^{-1}$. A dialysate-feed, immobilized-cell dialysis continuous ethanol process was studied [108]. The inherent advantages of removing ethanol inhibition was offset by relatively low productivity, which appeared to be limited by membrane permeability. Lee and Chang [109] showed that the yeast biomass in the reactor can be controlled using appropriate rate of cell bleeding and obtained a cell mass of 210 g l^{-1}. Membrane recycling was used to produce ethanol from whey using *Kluyveromyces fragilis* [110] and from xylose using *Pachysolen tannophilus* [111], but the productivities were not as high as with a glucose substrate.

Membrane cell recycling was used to produce lactic acid [112–117]. The productivity was very high and continuous production of lactic acid was possible.

Pierrot et al. [118], Schlote et al. [119] and Afscheu et al. [120] used hollow-fiber ultrafilters to separate and recycle cells in a continuous process of *Clostridium acetobutylicum* to increase the productivity of acetone-butanol production.

5.3.2 Aerobic Cultures

The cell recycle system is the same as that used in anaerobic cell culture, but the difficulty is how oxygen can be supplied to meet the demand of a high density cell mass. In other words, unless sufficient is supplied, the cell mass cannot be maintained. Holst et al. [121] produced superoxide dismutase (SOD) using this system. The final biomass of *Streptococcus lactis* was 19 g l^{-1} which was much higher than the 1.7 g l^{-1} of the batch system. In terms of total SOD productivity, the cell recycle system was 3.5 times more productive than the batch system. Enzminger and Asenjo [122] produced citric acid using *Saccharomyces lipolytica* in the cell recycle reactor. The reactor productivity was $1.16 \text{ g l}^{-1} \text{ h}^{-1}$. Lee et al. [123] used this system to obtain a high density of recombinant *E. coli* biomass 50 g l^{-1} for tryptophan production. This value is much higher than the 12 g l^{-1} of recombinant *E. coli* obtained by Anderson et al. [124]. The problem in the *E. coli* culture is that *E. coli* loses plasmids very rapidly and also leakage through the fibers was frequently observed. Lee and Chang [125] successfully produced high cell cultures of recombinant *E. coli* using cell recycle cultures to obtain cell densities of about 100 g l^{-1}. Acetic acid was produced with *Acetobacter* sp. [126]. High cell-density was realized by the integrated process and productivity was much higher than by batch culture.

5.3.3 Kinetics in Membrane Recycle Reactors

The main advantage of membrane cell recycle reactors is to run the reactor at a higher dilution rate without worrying about washout than with continuous reactors. This becomes possible by separating hydraulic retention time (HRT) and solid retention time (SRT). Figure 21 shows a schematic diagram of cell recycle

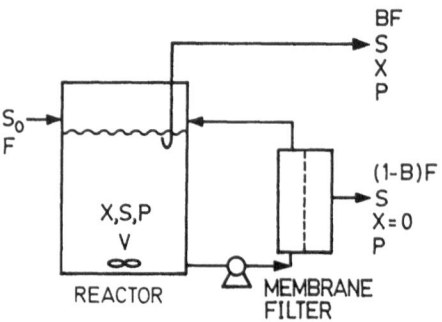

Fig. 21. Schematic diagram of membrane recycle system [109]: B, bleed ratio; S, glucose concentration; F, feed rate; X, cell concentration; P, product concentration; V, reaction volume

system. Writing material balances for S, P and X are

$$dX/dt = (\mu - BD)X,\tag{4}$$

$$\mu = \mu_{max}\frac{S}{S + K_s},\tag{5}$$

$$dS/dt = D(So - S) - \mu X/Yx/s,\tag{6}$$

$$dP/dt = -DP + q_pX.\tag{7}$$

At steady state $dX/dt = 0$ and the condition for no washout is

$$D_{washout} < \mu/B,\tag{8}$$

where B stands for bleed ratio. If there is no bleed, $D_{washout}$ becomes infinite. When the product is intracellular, the productivity (π) of a reactor is

$$\pi = \gamma\mu X,\tag{9}$$

where γ is the intracellular product concentration and if a product is extracellular, the product concentration P

$$P = q_pX/D\tag{10}$$

and the productivity π consists of two parts

$$\pi = (1 - B)DPa + BDP,\tag{11}$$

where the first term refers to the productivity obtained through membrane filtration and Pa is the product concentration through the membrane. The second term refers to a product that goes out with a bleed stream containing the cell mass. From Eq. (10) we can see that P becomes lower with the increase of dilution rate D. The same argument applies to an inhibitor or a toxin that inhibits cell growth or product formation if present. Probably this is the main reason that high density cell culture with a high productivity is possible in a membrane recycle system in comparison with batch or fed-batch systems.

Figure 22 shows an increase in *S. cerevisiae* cell concentration in a membrane cell recycle system. The cell is grown in a batch and then a cell recycle culture with no bleed begins. The cells cannot fully utilize the substrate supplied in excess and as a result, the substrate level increases for a while. The ethanol concentration decreases since P at steady state by Eq. (10) is lower than P accumulated in a batch system. However, as the time goes on, the biomass increases and the substrate is used up, the rate of biomass increase in the reactor is

$$dX/dt = \mu X = Yx/sDSo = constant.\tag{12}$$

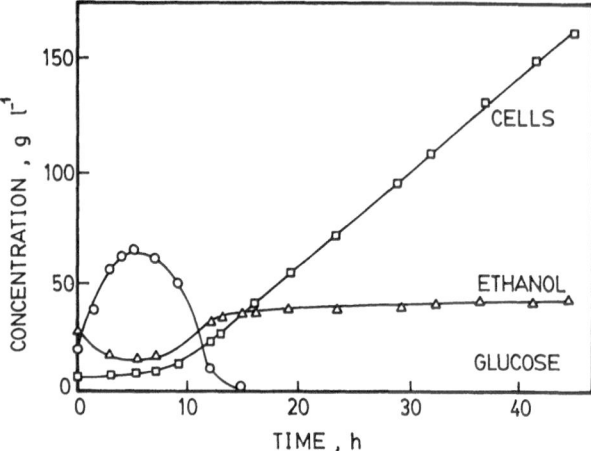

Fig. 22. Reaction kinetics in a membrane recycle reactor [109]: ethanol production with *S. cerevisiae*, So = 85 g l^{-1}, D = 0.36 h^{-1}, bleed ration = 0

Since this is a substrate-limited growth, Eq. (5) does not apply here. From the slope we can calculate Yx/s. Once the biomass reaches a level that can utilize all the substrate supplied continuously, the effluent glucose concentration drops to zero and ethanol concentration remains constant. However, the biomass concentration keeps increasing. Once the biomass reaches a desired level, say 150 g l^{-1}, we have to bypass a certain portion of the effluent stream. At steady state dX/dt = 0. The amount wasted is

$$\mu X = BDX. \tag{13}$$

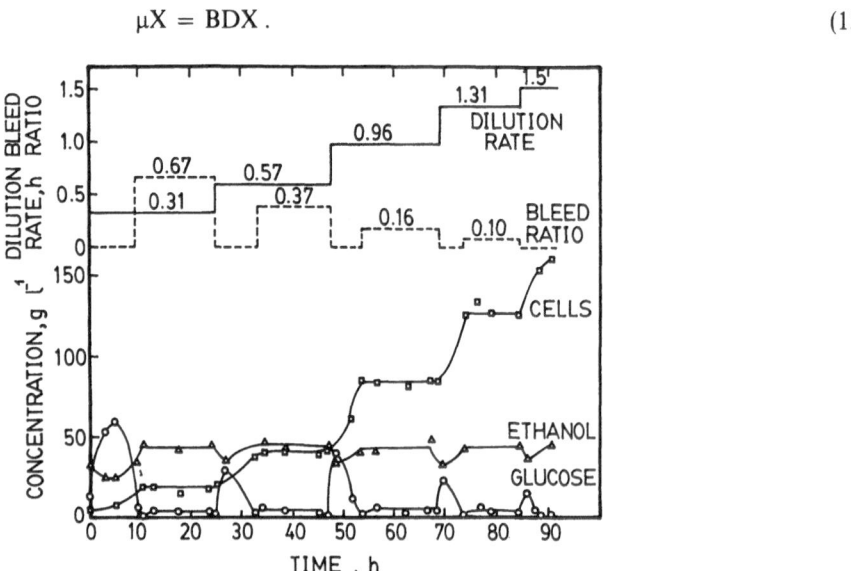

Fig. 23. Steady-state kinetics with bleed [109]

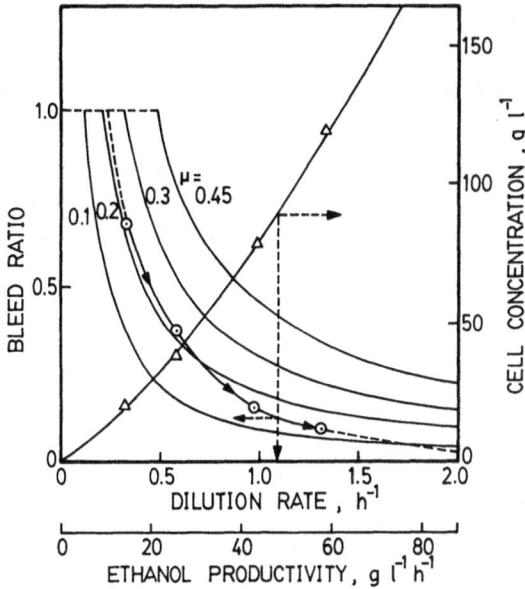

Fig. 24. Determination of bleed ratio as a function of dilution rate and cell concentration at So $= 100 \, \text{g} \, l^{-1}$ [109]

Figure 23 shows how to manipulate bleed vs dilution rate. At a high dilution rate the bleed ratio should be lower because of the relationship given in Eq. (13). If μ is constant, B and D are inversely proportional. However μ can vary with dilution rate and the actual relationship is represented as the trajectory shown in Fig. 24.

One of the difficulties associated with the cell recycle culture is how to recycle large numbers of cells. On a laboratory scale and with a clean substrate, a membrane system will work, but on a larger scale, a centrifugational cell separation system can be adopted. Anyway, developing an easy and economical cell recycle system is a problem to be solved for commercial use of this process. In the meantime the membrane cell recycle system will serve as a tool to study cell growth and product formation kinetics for high density cell culture systems. In addition to cell recycling, the supply of large amounts of oxygen and heat removal can also be a problem [127].

5.4 Prospects

The high volumetric productivity and high density cell mass in dual hollow fiber bioreactors and membrane recycle reactor systems are very promising in comparison with the conventional batch system. The dual hollow fiber bioreactor seems to be an ideal choice for aerobic production without having to worry about washout. In cell recycle reactor systems, high density culture of aerobic microorganisms becomes difficult because of the limitation in oxygen supply. However, anaerobic cell cultures can be carried out at high cell densities. The problems in recycle reactors are the plugging of hollow fibers and subsequent decrease of filtration flux. This problem can be alleviated by designing the distributor head properly [18].

6 Animal and Plant Cell Cultures

6.1 Characteristics of Animal Cell Cultures

Animal cells are very fragile and shear sensitive [128–133]. When the cells are grown in suspension, special precaution should be exercised to prevent shear-induced damage. Animal cells also require a lot less oxygen and the growth rate is lower than microbial cells (Table 8).

Table 8. Oxygen requirement of animal tissue cultures

	Animal cells	Microbial cells
Cell size	20–30 μm	1–3 μm
Oxygen req. (specific)	0.02–0.11 μmoles per 10^6 cells per h	3 mmole per g dry wt.
Typical cell concentration	10^6 ml^{-1} (10^8 ml^{-1})*	10 g l^{-1} (500 g l^{-1})*
Oxygen req. (volumetric)	0.02–0.11 mmole l^{-1} (2–11 mmole l^{-1})*	30 mmole l^{-1} (1500 mmole l^{-1})*

* Maximum cell densities

Anchorage-dependent cells require surfaces to grow on. Hollow fibers and membranes provide good surfaces for these cells. Because of this first animal cell immobilization system was carried out in hollow fibers. Anchorage-independent cells such as hybridomas do not require surfaces to attach to. They can be grown in suspension. In this case membranes are used to retain the cells to prevent washout.

6.2 Animal Cells Immobilized in Hollow Fibers

In 1972, Knazek et al. [4] cultured human choriocarcinoma cells on mixed bundles of Amicon XM-50 and Dow silicone polycarbonate capillaries. A Dow Corning Mini-lung was used to oxygenate the culture medium before feeding it into the capillaries. Human chorionic gonadotrophin (hCG) was produced for more than 28 d by the capillary culture, with a cell density of 217×10^6 cells in a 3 cm^3 culture space. A flat-bed hollow fiber cell culture system was developed by Ku et al. [134]. The air-CO_2 mixture was supplied to the cells via the lumen of the fibers. A significantly greater cell concentration of SV3T3 cells was obtained under pulse aeration (2.14×10^6 cells per cm^2) than under unaerated conditions (1.40×10^6 cells per cm^2). The hollow fiber system was operated for periods of 21 to 59 d. The data obtained in the flat-bed systems with 930 and 9300 cm^2 surface areas supported efficient utilization of the fiber surface and minimal gradient development. Human (β) interferon was produced in a culture with microcarrier beads added in the extracapillary space of Vitafiber 3S100 and 3P10 matrix perfusion culture units [135, 136]. Interferon did not traverse the 3P10 (10,000 Da porosity) membrane during on 8-h trial, while it readily diffused through the 3S100 fiber walls (100,000 Da porosity) during 8-h perfusion. Interferon yield upon induction in the perfusion system was more than 12 times that in a mono-layer culture in a flask. A radial flow hollow fiber bioreactor was used to grow H1 cells to a density of 7.3×10^6 cm^{-3} [137–140]. The reactor consisted of a central flow distributor tube surrounded by an annular bed of hollow fibers. A mixture of air and CO_2 is fed through the tube side of the hollow fibers. Spent medium diffused across the tube side of the hollow fibers and was convected away with the spent gas stream. A hydridoma culture was maintained over extended periods in polysulfone hollow fiber membrane modules where the molecular cutoffs were 10,000, 50,000 and 100,000 [141]. The system employed in this study was the same as that used by Knazek et al. [142]. The module with the 100,000 molecular weight cutoff membrane accumulated the highest level of IgG, but the leakage of IgG into the reservoir was unavoidable due to the large membrane pores.

6.3 Cells Retained by Membranes

Major products derived from animal cell culture are proteins. Thus one advantage of using membranes in animal cell culture is that size-selective membrane can screen out unnecessary impurities from products we want. Klement et al. [143] used double membranes to prevent mixing of serum proteins and product proteins (Fig. 25). In other cases membranes are used just for retaining the cells to prevent cell washout. So the special term of "perfusion culture" is used. Not only membranes of polymers but also stainless steel or nylon meshes can be used [144, 145]. Unlike in the membrane recycle cultures for microbial cells the cell densities achieved by perfusion system in suspension is about 10^7 cells per ml, which is only 1/10 of what can be achieved by matrix immobilization.

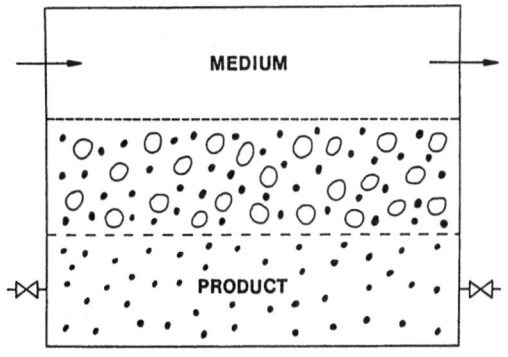

Fig. 25. Double membrane reactors for hybridoma growth [143]

6.4 Oxygen Transfer Systems

Oxygen is an important nutrient for cell viability and productivity of animal cells as in microbial cells. But its specific requirement per unit volume is only 1/100 or 1/1000 of those of microbial cells (Table 8). Thus bubbling or other means of aeration used in microbial bioprocess can supply enough oxygen to animal cell cultures. But the problem is that bubble aeration can cause excessive foam generation from media containing serum or proteins produced and produces shear damage in fragile animal cells. Surface and membrane aeration are commonly applied [134, 146]. The cell culture system developed by GBF employs a porous polypropylene tubing for oxygen supply [147]. The balance between the hydraulic pressure of the medium and the internal pressure of the tubing prevents air from bubbling through the medium.

Hollow fibers by themselves do not supply enough oxygen to aerobic microbial cells growing in the module. But animal cells can grow in the hollow fiber module as we have seen in Knazek's work [4]. This is due to the lower oxygen requirement of animal cells. The second important factor is that animal cells are weak and well retained among hollow fibers unlike in the case of *Aspergillus niger* for critic acid production [89]. But according to Ku et al. [134], it is evident that animal cells immobilized in hollow fibers are also short of oxygen [49]. Also the oxygen penetration depth estimated by Chang and Moo-Young is about 500–1000 µm for animal cells in hollow fibers. Since the fibers in hollow fiber modules are usually, further apart than this estimation, animal cells will be short of oxygen to some extent. The dual hollow fiber reactors applied to animal cell culture in the author's laboratory resulted in a significant improvement in monoclonal antibody production [148]. In fact a commercial system developed by Endotronics uses pressure swing for enhanced nutrient exchange including oxygen between lumen and shell side. Even though cell immobilization leads to limitation of nutrient transport, higher specific productivity of cells caused by cell to cell interaction is an advantage of large-scale animal cell cultures.

6.5 Commercial Animal Cell Culture Systems

Hollow fiber reactors are most widely used for small scale animal cell cultures. Amicon uses the original system developed by Knazek et al. [149]. Endotronics uses a pressure pulsing system to promote mass transfer. Bioresponse is a company that uses a membrane to retain cells and utilizes filtered cow lymph for providing the medium component. Recently Setec Co. developed a dual hollow fiber reactor, tricentric system similar to the natural blood vessel system. The problem with these systems is the difficulty of scaling up the process rather than stacking several hollow fibers in series.

Compared to other systems hollow fiber systems have advantages such as the operation with variable modes, the reactor stability, the usage of all cell types and the higher value of the surface to volume ratio (S/V) [150].

Monsanto uses a spin-filter system and NBS's CelliGen reactor uses a screened-in chamber enclosed by a 200-mesh stainless steel to retain foams from going outside [145, 151]. Damon Biotech's microencapsulation process can be used in large-scale culture system [152].

6.6 Plant Cell Culture Systems

Higher plants provide many valuable biochemicals including perfumes, dyes, medicinals, and opiates. Most of these chemicals are produced by cultivation of whole plants. Recent improvements in plant cell culture techniques have provided an alternative to whole plant cultivation, and there is a significant potential for future applications of plant cell cultures for the production of valuable plant-derived chemicals.

There are some reviews on plant cell immobilization methods and reactors [153–157]. Plant cells have been entrapped in gels of alginate [158, 159], carrageenan, polyacrylamide, agarose [160], polyurethane foam [161] and hollow fiber unit, or in the combined form of these materials. Hollow fiber membrane bioreactors have been used for plant cell cultures and secondary metabolite production [162–167]. The first reported use of a hollow fiber unit for the plant cell immobilization dealt with phenolics production by the cultures of *Glycine max* (soy bean) [162]. Kim et al. [168] used a dual hollow fiber membrane bioreactor to enhance phenolic productivity to about 58 times higher than that of a shake flask.

The flat membrane bioreactor system has been investigated by Shuler et al. [162, 169]. The membranes provide compartmentalization and organization. The hydrophobic membrane between the cell layer and the gas phase promotes oxygen supply and CO_2 removal from the cell layer. Another hydrophobic membrane separating culture medium and extractant is used to separate the aqueous nutrient phase from an organic solvent. The purpose of this solvent is to selectively absorb the product, and removal of an inhibitory product can greatly increase production rate and potentially reduce the recovery costs. A hydrophilic membrane is used to separate the cell layer from the nutrient solution. The cells are present in high concentration to achieve higher volumetric productivity. When membrane bioreactors are used in combination with the solvent extraction of plant products, these will be very useful reactors for plant cell cultures.

7 Concluding Remarks

Retaining biocatalysts in or with membranes was initiated about 30 years ago. Many of original concepts are still being studied for further developments. Immobilization of enzymes in hollow fiber reactors has widened its application from simple bioconversion to extractive bioconversion and production of amino acids using immobilized cofactors. Many chemicals that can be biologically synthesized will use membrane enzyme reactors for their production. Immobilizing microbial cells in the membrane matrix presents problems to be solved for further application since immobilized microorganisms tend to overgrow to break down hollow fibers. However, the membrane recycle reactor is a feasible system for high cell density cell culture if proper filtration systems are available so that industrial substrates can be used. If this system runs successfully, separation of hydraulic residence time and solid retention time becomes possible. Then wash-out problem will be eliminated in a continuous culture. Thus it is important to develop a membrane cell recycle system that can handle industrial substrates over a long period.

The use of membrane reactors for animal cell culture is one of the earliest achievements in commercial membrane bioreactor systems. Animal cells are fragile and do not require much oxygen in comparison with microbial cells. Furthermore it is a small scale process, and thus it accomodates all advantages and disadvantages of membrane bioreactors. Thus membrane bioreactors are considered to be ideal candidates for animal cell cultures.

Plant cells resemble animal cell cultures in many aspects, but the growth of plant cells in a small membrane space can cause the break-up of membranes such as those we experienced in fungal cell growth in dual hollow fiber systems. It is too early to say whether the membrane reactor is suitable for plant cell culture or not.

For further use of membrane bioreactors in biotechnology processing, new module design other than the currently available hollow fibers should be pursued. For instance membrane modules that can withstand high pressure should be developed and membrane materials other than polymers should be as easily fabricated as polymers.

8 Acknowledgements

The assistance of Dr. Dong Jin Kim in the preparation of this manuscript is appreciated. The first author (Ho Nam Chang) is grateful to the Korean Science and Engineering Foundation (KOSEF) and Korean Advanced Institute of Science and Technology (KAIST) for the research grants used in membrane bioreactors.

9 Symbols and Explanations

B	bleed ratio in membrane cell recycle culture
D	dilution rate (h^{-1})
K_S	Monod constant $(mol\, l^{-1})$
L	length of hollow fibers (cm)
n	number of hollow fibers in a hollow fiber module
P	product of a bioreactor $(g_p\, l^{-1})$
Q	total flow rate in a hollow fiber reactor
Qo_2	specific oxygen uptake rate per biomass $(mM\, g^{-1}\, h^{-1})$
q_p	specific biomass production rate $(g\, g^{-1}\, h^{-1})$

R	radius of hollow fibers (cm)
S_0	inlet substrate concentration (mol l^{-1})
S	substrate concentration in the reactor (mol l^{-1})
X	cell concentration in the reactor (g l^{-1})
$Y_X\Gamma_S$	biomass yield (biomass per substrate, g g^{-1})
α	coefficient related to membrane permeation
β	coefficient related to pressure drop without membrane permeation
γ	coefficient related to total reactor productivity
ΔP	pressure drop in a single hollow fiber
θ	time for oxygen depletion in hollow fibers (s)
μ	specific growth rate (h^{-1})
μ_{max}	maximum specific growth rate (h^{-1})
π	productivity (g l^{-1} h^{-1})
φ	volume fraction of hollow fiber or microsphere

10 References

1. Gallup DM, Gerhardt P (1963) Appl Microbiol 11: 506
2. Chang TMS (1964) Science 146: 524
3. Butterworth TA, Wang DIC, Sinskey AJ (1970) Biotechnol Bioeng 12: 615
4. Knazek RA, Gullimo PM, Kohler PO, Dedrick RL (1972) Science 178: 655
5. Kitano H, Ise N (1984) Trends in Biotech 2: 5
6. Gekas VC (1986) Enz Microb Tech 8: 450
7. Vick Roy TB, Blanch HW, Wilke CR (1983) Trends in Biotech 1: 135
8. Hopkinson J (1983) Hollow fiber cell culture in industry. In: Mattiasson B (ed) Immobilized cells and organelles, vol I, CRC, Boca Raton, FL, p 89
9. Belfort G (1989) Biotechnol Bioeng 33: 1047
10. Karel SF, Libicki SB, Robertson CR (1985) Chem Eng Sci 40: 1321
11. Furusaki S (1988) J Chem Eng Japan 21: 219
12. Chang HN (1987) Biotech Adv 5: 129
13. Eisenman G (1972), Membranes vol 1 Macroscopic systems and models. Marcel Dekker, New York, p 3
14. Atkinson B, Mavituna F (1983) Biochemical and biotechnology handbook. McMillan, Surrey, England, p 933
15. Kesting RE (1985) Synthetic polymeric Membranes, 2nd ed John Wiley, New York
16. Belfort G (1988) J Membrane Sci 35: 245
17. Chang HN, Park JK (1986) Effect of turbulence promoters in mass transfer In: Cheremisnoff NP (ed) Handbook of heat and mass transfer operations vol II, Gulf Publishing, USA, p 3
18. Park JK, Chang HN (1986) AIChE J 32: 1937
19. Lightfoot EN (1974) Transport phenomena and living systems, John Wiley, New York p 99
20. Waterland LW, Robertson CR, Michaels AS (1975) Chem Eng Comm 2, 37
21. Kang IS, Chang HN (1982) Int J Heat Mass Transfer 25: 1167
22. Kim DH, Chang HN (1983) Int J Heat Mass Transfer 26: 1007
23. Kim WS, Park JK, Chang HN (1987) Int J Heat Mass Transfer 30: 1183
24. Lee YL, Chang HN (1987) J Korean Inst Chem Engr 26: 97
25. Gabler FR (1985) In: Moo-Young M (ed) Comprehensive biotechnology vol II Pergamon Press, Oxford 111, p 351
26. Furusaki S, Kojima T, Miyauchi T (1977) J Chem Eng Japan 10: 233
27. Kim IH, Chang HN (1983) AIChE J 29: 645

28. Kim IH, Chang HN (1983) AIChE J 29: 910
29. Park TH, Kim IH, Chang HN (1985) Biotechnol Bioeng 27: 1185
30. Chang HN, Chung BH, Kim IH (1986) In: Asenjo JA, Hong J (eds) Separation, recovery and purification in biotechnology: Recent advances and mathematical modeling, ACS Symposium Series 314: 32
31. Smith WJ, Salmon PM, Robertson CR (1988) ACS 196th meeting, MBTD paper 44
32. Chang HN, Kyung YS, Chung BH (1987) Biotechnol Bioeng 29: 552
33. Chang HN, Chung BH (1987) Korean J Chem Eng 5: 83
34. Uttapap D, Koba Y, Ishizaki A (1989) Biotechnol Bioeng 33: 542
35. Darnoko D, Cherylan M, Artz WE (1989) Enz Microb Technol 11: 154
36. Hong J, Tsao GT, Wankat PC (1981) Biotechnol Bioeng 23: 1501
37. Pizzichini M, Fabiani C, Adami A, Cavazzoni V (1989) Biotechnol Bioeng 33: 955
38. Deeslie WD, Cherylan M (1982) Biotechnol Bioeng 23: 2257
39. Deeslie WD, Cherylan M (1982) Biotechnol Bioeng 24: 69
40. Bressollier Ph, Petit JM, Julien R (1988) Biotechnol Bioeng 31: 650
41. Kroner KH, Nissinen V, Ziegler H (1989) Bio/Technology 5: 921
42. Horvath C, Solomon BA (1972) Biotechnol Bioeng 14: 885
43. Drioli E, Gianfreda L, Palescandol R, Scardi V (1975) Biotechnol Bioeng 17: 1365
44. Kawakami K, Harada T, Kusunoki K (1980) Enz Microb Technol 2: 295
45. Wichmann R, Wandrey C, Buckmann AF, Kula M-R (1981) Biotechnol Bioeng 23: 2789
46. Berke W, Schulz H-J, Wandrey C, Morr M, Denda G, Kula M-R (1988) Biotechnol Bioeng 32: 130
47. Ishikawa H, Takase S, Tanaka T, Hikita H (1989) Biotechnol Bioeng 34: 369
48. Miyawaki O, Nakamura K, Yano T (1982) J Chem Eng Japan 15: 224
49. Chang HN, Moo-Young M (1988) Appl Microb Biotechnol 29: 107
50. Chang HN, Joo IS, Ghim YS (1984) Biotechnol Lett 6: 487
51. Hoq MM, Yamane T, Shimizu S, Funata T, Ishida S (1984) JOACS 61: 776
52. Pronk W, Kerkhof PJAM, van Helden C, van't Riet K (1988) Biotechnol Bioeng 32, 512
53. Bratzler RL (1987) In: Biotransformation, The World Biotech Report vol 1, part 5, p 23
54. Ishikawa H, Kurose K, Oogaito M, Hikita H (1989) J Chem Eng Japan 22: 18
55. Shimazaki K (1987) In: Current chemical engineering (Bioreactors accompanying separation) vol 39, p 16
56. Imai K, Shiomi T, Uchida K, Miya M (1986) Biotechnol Bioeng 28: 198
57. Shiomi T, Tohyama M, Satoh M, Miya M, Imai K (1988) Biotechnol Bioeng 32: 664
58. Imai K, Shiomi T, Uchida K, Miya M (1986) Biotechnol Bioeng 28: 1721
59. Furusaki S, Nozawa T, Nomura S (1990) Bioprocess Eng 5: 73
60. Lee CK, Hong J (1988) Biotechnol Bioeng 32: 647
61. Kunugi S, Kodama H, Yamada H, Nakamura Y (1985) Sen'i Gakkaishi 41: T355
62. Ichijo H, Suehiro T, Nagasawa J, Yamauchi A, Sugesaka M (1985) Biotechnol Bioeng 27: 1077
63. Gianfreda L, Domenico P, Greco Jr G (1989) Biotechnol Bioeng 33: 1067
64. Chung BH, Chang HN, Koh YH (1987) J Ferment Tech 65: 575
65. Hwang JS, Chang HN (1987) Biotechnol Lett 9: 237
66. Leuchtenberger W, Karrenbauer M, Plocker U (1983) Enzyme Engineering 7, Annals of NY Acad Sci 434: 78
67. Tanigaki M, Sakata M, Wada H (1987) In: Bioreactors functioning separation, Soc Chem Japan (Latest Chemical Eng) 39: 33
68. Hibino T (1987) In: Bioreactors functioning separation, Soc Chem Eng Japan (Latest Chemical Eng) 39: 33
69. Hawashida M, Kise S, Ikemi M, Koizumi S (1988) In: Biotechnology (Prep Sixth Symp Fundam Technol for Next Generation) p 155
70. Wang HY, Kominet LA, Jost JL (1981) In: Moo-Young et al (eds), Advances in biotechnology vol I, Pergamon Press, Oxford, p 601
71. Schultz JS, Gerhardt P (1969) Bacterial Rev Mar p 1
72. Matsumura M, Markl H (1986) Biotechnol Bioeng 28: 534
73. Matsumoto K, Ohya H, Daigo M (1985) Membrane 10: 305

74. Matsumoto, Ohya H (1986) World 3 Cong Chem Eng Sept 21–25, p 843
75. Nakao S, Saitoh F, Asakura T, Toda K, Kimura S (1987) J Memb Sci 30: 273
76. Udriot H, Ampuero S, Marison IW, von Stokar U (1989) Biotech Lett 11: 509
77. Vick Roy TB, Blanch HW, Wilke CR (1982) Biotech Lett 4 (8): 483
78. Inloes DS, Taylor DP, Cohen SN, Michaels AS, Robertson CR (1983) Appl Environ Microbiol 46: 264
79. Mehaia MA, Cherylan M (1984) Appl Microbiol Biotechnol 20: 100
80. Mehaia MA, Cherylan M (1984) Enz Microbiol Technol 6: 117
81. Park TH, Kim IH (1985) Appl Microbiol Biotechnol 22: 190
82. Ooshima H, Dono N, Harano Y (1987) Prep 52nd Annual Meeting Soc Chem Japan, Nagoya p 348
83. Jee HS, Nishio N, Nagai S (1988) Biotech Lett 10: 243
84. Mattiasson B, Ramstorp M (1981) Biotech Lett 3: 561
85. Inloes DS, Smith WJ, Taylor DP, Cohen SN, Michaels AS, Robertson CR (1983) Biotechnol Bioeng 25: 2653
86. Nanba, Kimura K, Nagai S (1985) J Ferment Technol 63: 175
87. Robertson CR, Kim IH (1985) Biotechnol Bioeng 27: 1012
88. Chang HN, Chung BH, Kim IH (1986) In: Asenjo JA, Hong J (eds) Recent advances in purification and mathematical modelling, ACS Symposium Series 314, 32–42
89. Chung BH, Chang HN (1988) Biotechnol Bioeng 32: 205
90. Kristiansen B, Sinclair CG (1978) Biotechnol Bioeng 20: 1711
91. Anderson JG, Blain JA, Divers M, Todd JR (1980) Biotechnol Lett 2: 99
92. Vaija J, Linko YY, Linko P (1982) Appl Biochem Biotechnol 7: 51
93. Eikmeier H, Rehm HJ (1984) Appl Microbiol Biotechnol 20: 365
94. Horitsu H, Adachi S, Takahashi Y, Kawai K, Kawano Y (1985) Appl Microbiol Biotechnol 22: 8
95. Tsay SS, To KY (1987) Biotechnol Bioeng 29: 297
96. Chung BH, Chang HN, Kim IH (1987) Enz Microbiol Tech 9: 345
97. Hwang YB, Chung BH, Chang HN, Han MH (1988) Bioprocess Eng 3: 159
98. Chung BH, Chang HN (1985) Korean J Appl Microb Bioeng 13: 209
99. Chung BH, Cho DC, Chang HN (1987) Korean J Appl Microb Bioeng 15: 49
100. Herbert D (1961) Society of Chemistry and Industry Monograph No 12, London, p 21
101. Pirt SJ, Kuroski WM (1970) J Gen Microbiol 63: 357
102. Margaritis A, Wilke CR (1978) Biotechnol Bioeng 20: 727
103. Rogers PL, Lee KJ, Tribe DE (1980) Process Biochem 15: 7
104. Nishizawa Y, Mitani Y, Tamai M, Nagai S (1983) J Ferment Technol 61: 599
105. Nishizawa Y, Mitani Y, Tamai M, Nagai S (1983) J Ferment Technol 61: 599
106. Nishizawa Y, Mitani Y, Fukunishi K, Nagai S (1984) J Ferment Technol 62: 41
107. Cherylan M, Mehaia MA (1984) Process Biochem December 204
108. Kyung KH, Gerhardt P (1984) Biotechnol Bioeng 26: 252
109. Lee CW, Chang HN (1987) Biotechnol Bioeng 29: 1105
110. Janssens JH, Bernard A, Bailey RB (1984) Biotechnol Bioeng 26: 1
111. Chung IS, Lee YY (1986) Biotechnol Bioeng Symp 15: 249
112. Vick Roy TB, Mandel DK, Dea DK, Blanch HW, Wilke CR (1983) Biotechnol Lett 5: 665
113. Kobayashi T, Minami T (1988) Kemikaru Enjiniaringu 33: 965
114. Tachigawa S, Seike Y, Inaba H, Nagamune T, Endo I (1988) Kemikaru Enginiaringu 33: 970
115. Taniguchi M, Kotani N, Kobayashi T (1987) J Ferment Technol 65: 179
116. Taniguchi M, Hoshino K, Shimizu K, Nakagawa I, Takahashi Y, Fujii M (1988) J Ferment Technol 66: 633
117. Wang E, Hatanaka H, Iijima S, Takebayashi T, Shimizu K, Matsubara M, Kobayashi T (1988) J Chem Eng Japan 21: 36
118. Pierrot P, Fick M, Engasser JM (1986) Biotechnol Lett 8: 253
119. Schlote D, Gottschalk G (1986) Appl Microb Biotechnol 24: 1
120. Afscheu AS, Biebl H, Schaller K, Schügerl K (1985) Appl Microbiol Biotechnol 22: 394

121. Holst O, Hansson L, Berg AC, Mattiasson B (1985) Appl Microbiol Biotechnol 23: 10
122. Enzminger JD, Asenjo JA (1986) Biotechnol Lett 8: 7
123. Lee CW, Gu MB, Chang HN, (1989) Enz Microb Tech 11: 49
124. Anderson KW, Grulke E, Gerhardt P (1984) Bio/Technology 2: 891
125. Lee YL, Chang HN (1990) Biotechnol Bioeng (in press)
126. Park YS, Ohtake H, Fukaya M, Okumura H, Kawamura Y, Toda K (1989) Biotechnol
     Bioeng 33: 918
127. Goma G, Durand G (1988) 8th Int Biotechnol Symp Proc vol 1, p 410
128. Chittur KK, McIntire LV, Rich RD (1988) Biotechnol Progress 4: 89
129. Eppenberger RM (1988) J Biotechnol 7: 179
130. Papoutsakis ET (1988) J Biotechnol 7: 229
131. Palsson BO (1988) Biotech Lett 10: 625
132. Randerson DH (1987) Biotech Lett 1: 39
133. Dodge TC, Hu WS (1987) Biotech Lett 8: 683
134. Ku K, Kuo MJ, Delente J, Wildi BS, Feder J (1981) Biotechnol Bioeng 23: 79
135. Strand JM, Quarles JM, McConnel S (1984) Biotechnol Bioeng 26: 503
136. Strand JM, Quarles JM, McConnel S (1984) Biotechnol Bioeng 26: 508
137. Tharakan JP, Chau PC (1986) Biotechnol Bioeng 28: 329
138. Tharakan JP, Chau PC (1986) Biotechnol Bioeng 28: 1064
139. Tharakan JP, Chau PC (1986) Biotechnol Bioeng 29: 657
140. Gallager SL, Tharakan JT, Chau PC (1987) Biotechnol Techniq 1: 91
141. Altshuler GL, Dziewulski DM, Sowek JA, Belfort G (1986) Biotechnol Bioeng 28: 646
142. Gullino PM, Knazek RA (1979) Methods in Enzymol 58: 178
143. Klement G, Scheiter W, Katinger HWD (1988) In: Moo-Young (ed), Bioreactor
     immobilized enzymes and cells, Elsevier Applied Science, London, p 53
144. Himmelfarb P, Thayer PS, Martin HE (1969) Science 2: 555
145. Martin N, Brennan A, Denome L, Shaevitz J (1987) Bio/Technol 5: 838
146. Knazek RA, Kohler PO, Gullimo PM (1974) Experimental Cell Res 84: 251
147. Wagner R, Lehmann J (1988) Trends in Biotechnol 6: 101
148. Chang HN, Oh DJ (1988) ACS Los Angeles meeting, MBTD paper 46
149. Hopkinson J (1985) Bio/Technol 3: 225
150. Sinskey AJ (1983) Trends in Biotech 1: 534
151. Tolbert WR (1981) In Vitro 17: 885
152. Lim F (1984) Biomedical application of microencapsulations, CRC Press, Boca Raton,
     Florida, p 137
153. Brodelius P, Mosbach K (1982) J Chem Tech Biotech 32: 330
154. Prenosil JE, Pedersen H (1983) Enz Microb Tech 5: 323
155. Rhodes MJC (1985). IN Wiseman A (ed), Topics in enzyme and fermentation
     biotechnology, vol 10. Ellis Horwood, Chichester, p 51
156. Kargi F, Rosenberg MZ (1987) Biotech. Progress 3: 1
157. Panda AK, Mishira S, Bisaria VS, Bhojwani SS (1989) Enz Microb Tech 11: 386
158. Brodelius P, Deus B, Mosbach K, Zenk MH (1979) FEBS Lett 103: 93
159. Alfermann A, Schuller AW, Reinhard E (1980) Planta Med 40: 218
160. Brodelius P, Nilsson K (1980) FEBS Lett 122: 312
161. Lindsey K, Yeoman MM, Black GM, Mavituna F (1983) FEBS Lett 155: 143
162. Shuler ML (1981) Ann NY Acad Sci 369: 65
163. Shuler ML, Sahai OP, Hallsby GA (1983) Ann NY Acad Sci 413: 373
164. Jose W, Pedersen H, Chin C (1983) Ann NY Acad Sci 413: 409
165. Hallsby GA, Shuler ML (1986) Biotech Bioeng Symp 17: 741
166. Shuler ML, Hallsby GA, Pyne JW, Cho T (1986) Ann NY Acad Sci 469: 270
167. Cho T, Shuler ML (1986) Biotech Progress 2: 53
168. Kim DJ, Chang HN, Liu JR (1989) Biotech Techniq 3: 139
169. Shuler ML, Pyne JW, Hallsby GA (1984) JOACS 61: 1724

Shear Effects on Suspended Cells

Jose Celman Merchuk*
Department of Chemical Engineering, Ben Gurion University of the Negev, Beer Sheva, 84105, Israel

Shear has been mainly considered in the technical literature as a destructive element, when applied to microorganisms and cells. Indeed, most of the research work addressing the subject aims at the identification of damaging levels of shear on a given culture. The present work is focused on the effects of shear on suspended cultures before the damaging levels are attained. Inspection of the literature reveals that shear may influence growth rate, cellular volume, metabolite production rate and distribution, and membrane permeabilities. Available devices for study and evaluation of shear effects on suspended cultures are described and critically reviewed.

The review reveals the possibility of an influence of the liquid dynamics on the kinetics of the biochemical process. This is relevant for bioreactor design and scale up, and stresses the importance of using structural bioreactor models in order to describe the hydrodynamics of the system.

---

* This review is based on lectures presented by the author at the University College, London, on July 1988

Advances in Biochemical Engineering
Biotechnology. Vol. 44
Managing Editor: A. Fiechter
© Springer-Verlag Berlin Heidelberg 1991

1 Introduction

1.1 Shear in Bioreactors

The reactor scale-up in chemical processes is basically focused on obtaining in the large-scale vessel the same average rates of production as in a smaller-scale experimental reactor. It is not a simple task to reach this objective, since the hydrodynamics of large vessels are often complicated and difficult to model. The importance of the hydrodynamics stems from the fact that in large vessels, the transport of mass and heat are dependent on convective mechanisms, and are associated therefore to turbulent eddy flow. This implies the existence of adjacent parallel streams of fluid, each flowing with a different velocity. The spatial gradient of velocities between two streams is called shear rate.

$$y = \frac{dv_x}{dy} \tag{1}$$

where v_x is the velocity in the x direction and y is perpendicular to x. The shear rate can be seen as responsible for the convective transport and therefore both the mass transfer coefficient and the heat transfer coefficient will depend strongly on it.

Let us remember that for many liquids, conspicuously water, the shear rate y is directly proportional to the force per unit area or stress needed in order to maintain the difference of velocities (Newtonian liquids). This is called the shear stress τ, and the proportionality constant is the viscosity μ.

$$\tau = \mu\gamma \tag{2}$$

A review on models on non-Newtonian liquids and on methods and apparatus for rheological determinations is given by Atkinson and Mavituna [1].

Strictly speaking, Eqs. (1) and (2) apply only to laminar flow, and for turbulent flow a different "turbulent viscosity", ε, can be defined. The difference between μ and ε is that μ depends only on liquid properties and ε may depend also on the flow patterns.

In most chemical processes, the shear rate is not important by itself, but as a means of increasing the mass and heat transfer rates. In biochemical processes, on the other hand, the shear rate may be important by itself. This is certainly accepted for animal cells. In a recent publication, Bliem and Katinger [2] identify shear as one of the two phenomena recognized as critical for scale-up of animal cell culture processes. This can be extended to cell plant culture as well. It will be shown later that the technical literature gives testimony that also microbial cultures may be influenced by the effects of shear.

It is customary to admit that excessive shear can damage suspended cells, leading to their loss of viability and even disruption. In some cases, however, positive effects have been observed, within certain limits. It is clear that such positive effects may be due to enhancement of heat and mass transfer rates. It is very interesting to explore whether shear stress per se can have a beneficial effect on culture growth and metabolite production rates. This is done in the present paper by reviewing published data on the subject.

1.2 Scale-Up of Bioreactors

The problems encountered in the scale-up of bioreactors can be concentrated into two groups. One is covering the cases where a high power input per unit volume is used in the laboratory scale, but cannot be maintained in the industrial scale, due to economic or mechanic limitations. This is not the case of plant or animal cultures, where, in any case, very high specific power input cannot be used because of cell fragility. The other group of problems can be generalized as lack of knowledge about the hydrodynamics of large-volume vessels.

The methods available for scale-up of bioreactors have been reviewed by Osterhuis [3, 4], Kossen and Oosterhuis [5], and Sweere et al. [6]. Since in general the design from first principles can not be done because of lack of basic knowledge on the hydrodynamics of the bioreactors, one possible outlet is the "semifundamental method", consisting of using approximate simple models for the fluid dynamics, and integrating them with basic and known kinetics and heat and mass transfer rates.

It is not very common that all this information is available to a degree of certainty that allows safe design of a large system. Usually the designer must resort to dimensional analysis. This method requires the knowledge of all the variables affecting the process, which can be obtained from a qualitative, but realistic model [7].

A simplified version of this method is to limit the variables to one or two, and follow "rules of thrumb" [8], which, depending on the specific case, can be constant P/V, constant $k_L a$, etc. The literature shows, however, many cases of inconsistence of this method. For example, design of a scaled-up bioreactor keeping constant the oxygen transfer rate can lead to a better performance than expected, as in the case reported by Taguchi et al. [9] for gluco-amylase production by *Endomyces* sp., or worst than expected as for protease production by *Streptomyces* sp. [10]. Both results are shown in Fig. 1.

The method of regime analysis and scale down, proposed by Kossen and Oosterhuis [5], combines two tools in order to overcome the problem posed by the complexity of the biochemical reactors. It is based in considering the regime of the process in full scale as the objective, and from this point of view, plan the strategy of the process development.

It may be worthwhile to remember at this stage the basic ideas of regime analysis. Generally, biochemical processes involve a series of steps, some of them of mass or heat transfer by convection, some by diffusive mechanisms (activated or not), and some others where chemical reactions occur. In this last case, a mass transfer mechanism is superimposed, since molecules must encounter each other in order to react, and usually a heat effect will accompany the reaction. Depending on whether these steps are in parallel or in series, and on the relaxation time of each step [11], it often happens that the rate of the total process is given by the rate of one single step. But the equilibrium between all the individual rates can be (and usually is) upset by a change in scale. This is to be expected, since a change in scale will not bring a change

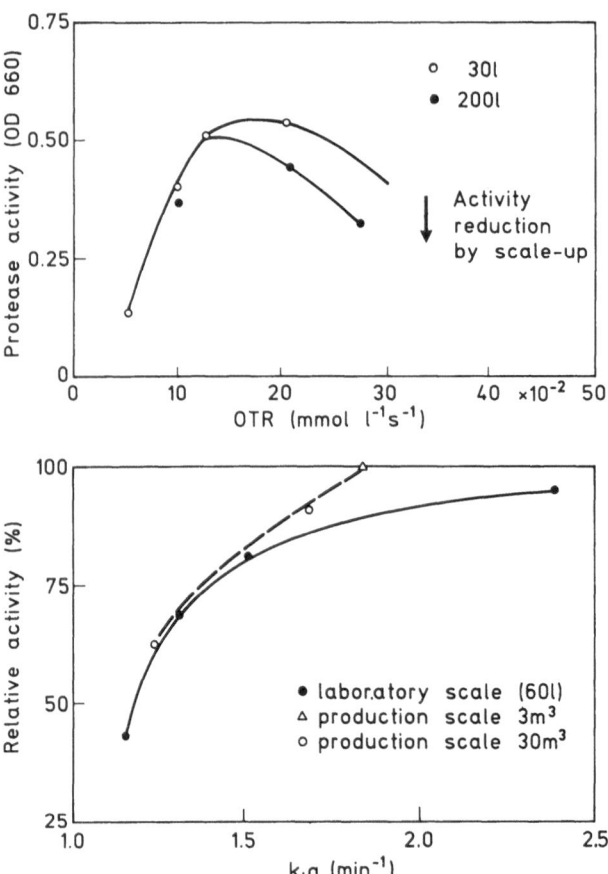

Fig. 1a, b. Opposite effect of scale-up at constant oxygen transfer rate in two different systems: Relative activity of gluco-amylase produced by *Endomyces sp.* reported by Taguchi et al. [9] and protease production by *Streptomyces sp.* reported by Takei et al. [10]

the physico-chemical or kinetic parameters (scale insensitive variables), but will affect the overall convective mass and heat rates (scale sensitive variables). A new equilibrium is established, and the interplay of all the parameters of the system may lead to a regime where a different step is controlling the process rate.

Oosterhuis-Kossen's [5] method proposes to start by an analysis of the operation of the large scale system. Once the regime is clarified, a small scale system is designed in such a way that it simulates the operation regime of the larger one. Optimization studies can be done on the smaller model, which will then be transferred to the full scale process. This is depicted in Fig. 2. An example of this method is given by Oosterhuis [4] where a large scale stirred tank reactor is simulated on a laboratory scale by two interconnected vessels, one with a small liquid volume, high agitation rate and high rate of oxygen supply (representing the zone near the impeller-sparger in the reactor), and a second vessel, with a much larger liquid volume, oxygen consumption (simulated by nitrogen sparging) and gentle agitation.

Production scale

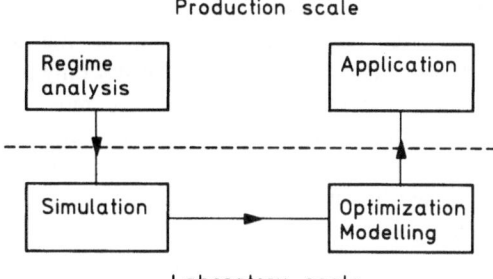

Laboratory scale   **Fig. 2.** The scale-down method [6].

The success of this model suggests that large vessels must be carefully analyzed and their internal structure studied in order to be properly modeled and designed. It is obvious that large vessels have not uniform hydrodynamic conditions in the whole of their volume and it makes sense to "lump" this volume into regions of easily definable velocity and turbulence. Each of·these regions constitutes a different environment, which may show different substrate and *product* concentrations, different pH, different cell concentration, different temperatures. Once this is defined, the equations representing the system can be written, since the kinetics of the reactions and the transfer rates depend on these variables.

The recognition of these different environments in the reactor in fact implies structured biokinetic models, but to the structure of the bioreactor, and this is generally a judgement on the hydrodynamics. This type of models can be very successful (see, for example, Bajpai and Reuss [12], Gibbs et al. [13]).

The main objective of this paper is to show that, in addition to the influence of the hydrodynamic conditions in each environment on the kinetics and activated diffusive

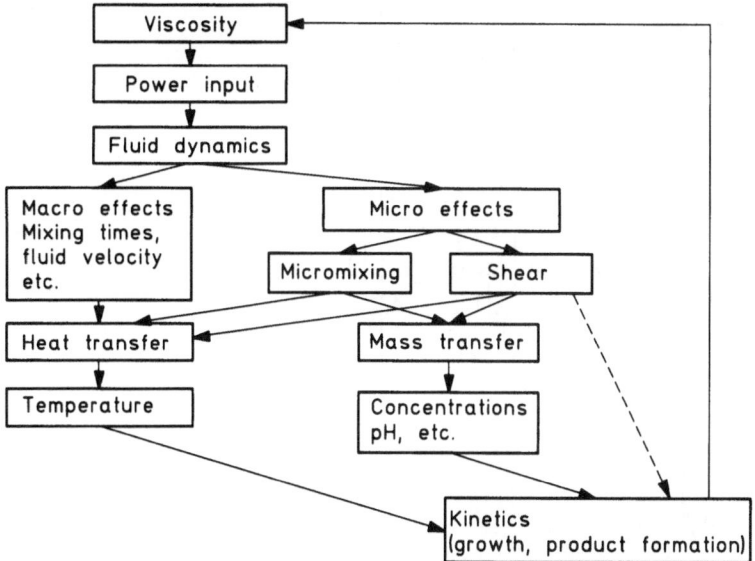

Fig. 3. Interactions between fluid dynamics and kinetics in bioreactors

processes (due to temperature and concentrations) and on the mass and heat transport rates (via the hydrodynamic-dependent transfer coefficients), the designer has to take into account the different shear conditions in each region. And this not only for the obvious reason that damaging conditions to the suspended culture must be avoided, but also because shear, even in pre-lethal levels, may affect the kinetics of growth and metabolic production, and must therefore enter as a new variable in the kinetic equations. This is shown by the broken line in Fig. 3 representing the inter-relation between viscosity and kinetics in a bioreactor.

If the above mentioned argument is accepted, the almost total lack of information in this area becomes evident. This paper reviews the available data and indications in the literature, for microorganisms, animal cells and plant cells, and describes the methods which have been proposed for the evaluation of shear effects on growth and metabolic production in suspended cultures.

2 Fluid Dynamics in Agitated Vessels

2.1 Characteristic Shear in a Stirred Tank

In spite of the remarkable quantity of novel designs for bioreactors [14–17] the conventional stirred tank is still the undisputed monarch in this kingdom. Data are available for liquid velocity distributions in standard configurations (Fig. 4). The

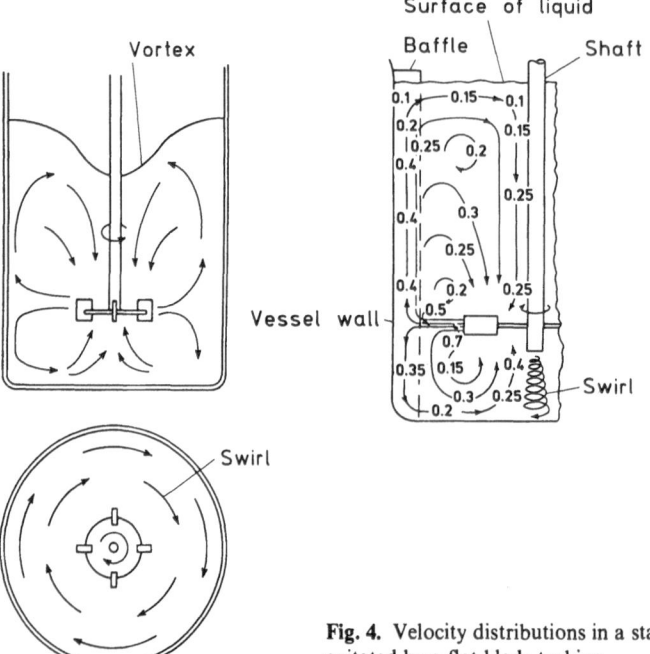

Fig. 4. Velocity distributions in a standard, baffled stirred tank agitated by a flat blade turbine

agitation is provided by an impeller, which is the channel for introduction of power into the system. Theoretical approaches to the agitation mechanics basically treat the impeller as a pump. The power input is divided into two parts: one is converted into the kinetic energy of the fluid and is responsible for the macroscopic flow that can be observed and measured. The other part is called head, which in turn encloses the static head or pressure and the viscous shear. The fraction of the power converted into flow which in turn is converted into shear, depends mainly on the type of the impeller. Some impellers will transform most of the power input into flow related energy, but without producing much mixing in the microscopic scale since not much velocity gradients are generated. Those impellers will produce a very gentle agitation, but micromixing and heat and mass transfer in the microscale will be very slow. On the other hand, an agitation method that converts most of the energy input into shear, will lead to highly efficient transfer rates in the microscale. In most applications of stirred tanks, both flow and shear are required in certain proportions and the selection of the type of impeller will define how the power input will be divided between these two main components. Fig. 5 shows a qualitative representation of this point given by Oldshue [18].

The quantitative determination of shear rates implies the measurement of flow profiles in the fermentor in order to calculate from them the velocity gradient, Eq. (1). This can be done by measurement of the fluid velocity in many points in order to map the whole tank. An example of such mapping can be seen on Fig. 6, taken from Weetman et al. [19]. Many anemometric techniques, from hot wire to laser doppler, have been used [18].

Since the energy enters focally through the impeller, high velocities and high velocity gradients will exist near the impeller, and they will decrease as they get closer to the walls. It should be kept in mind however, that at a given point the velocity may fluctuate in time, especially if — as in most applications — the flow is turbulent. The measured values of velocity are mean over a period of time.

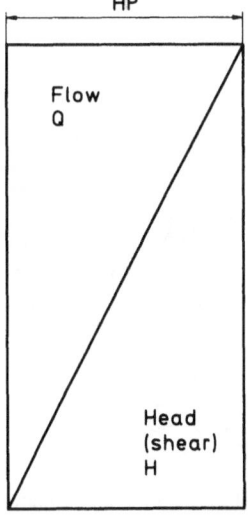

Rakes, Gates
Spiral, Anchors
Paddles
Propeller
Axial flow turbines
Flat blade turbines
Bar turbine
Bladeless impeller,
Impeller and stator
(close clearance)

Colloid mills
Homogenizer

Fig. 5. Partition of power input into stirred tanks for different impellers [18]

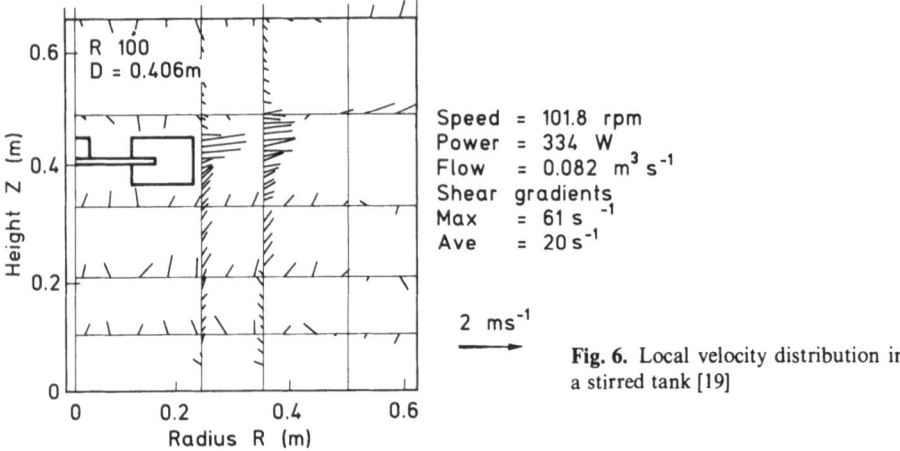

Speed = 101.8 rpm
Power = 334 W
Flow = 0.082 m³ s⁻¹
Shear gradients
Max = 61 s⁻¹
Ave = 20 s⁻¹

2 ms⁻¹

Fig. 6. Local velocity distribution in a stirred tank [19]

The measuring probe tests a given volume, and therefore the output is an average both in time and in space. That means that even if smooth curves are obtained, as those shown in Fig. 7 [18], in reality fluctuations may happen. This may be important when calculating the shear rates from the gradients of graphs as in Fig. 6, since velocities are obtained by an averaging procedure, and gradients larger than those detected may occur in the micro-scale. While the macro-scale shear rates operate on particles of 500 μm or larger, turbulent micro-scale shear rates operate on particles of 200 μm or less [20].

On the basis of the velocity profiles near the tip of the propeller, maximum and average shear rates in this region can be defined and calculated. A typical example is given in Fig. 8.

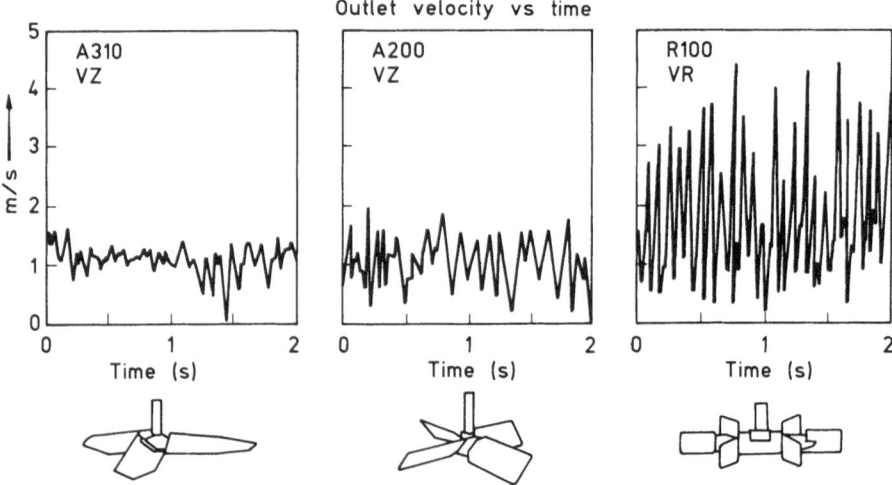

Fig. 7. Fluid velocities near the tip of impellers at different agitation speeds [19]

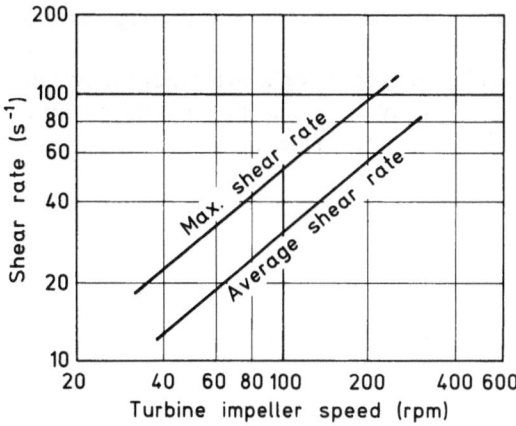

Fig. 8. Maximum and average shear rates in a 6-in six-blade turbine impeller rotating in a 18-in tank containing water [20]

In the surroundings of the impeller, the shear rates can be 50 to 100 times larger than those in other areas of the vessel. In addition to that, the root mean square fluctuation of the velocity can reach 50% of the mean measured velocity. This is obviously undesirable and stresses the need of bioreactors with an homogeneous shear rate field as well as the need to consider the fluid dynamic structure of bioreactors in modeling and simulation.

2.1 Characteristic Shear in a Stirred Tank

It is very useful that the shear in all the bioreactor can be represented by a single parameter, which characterizes its fluid dynamic behavior. If this mean or characteristic shear is correlated with geometrical and operational variables, a useful method for design is available. The first general expression in this matter was proposed by Metzner and Otto [21]. They proposed:

$$\gamma_{av} = kN \tag{3}$$

where k is a constant which depends only on the impeller geometry. Calderbank and Moo Young (22) extended the validity of Eq. (3) to non-Newtonian power low fluids, giving the following expression for k:

$$k = \left[\frac{4n}{3n + 1} \right]^{n(1-n)-1} B \tag{4}$$

where n is the exponent in Eq. (5), which describes the rheological behavior of pseudo-plastic fluids:

$$\tau = K\gamma^n \tag{5}$$

and B is a constant that they found to be 11.

An extension of this work to consider viscoelastic behavior with long relaxation times was done by Ducla et al. [23].

Stein [24] gives a simple mathematical model, based on a two zones stirred tank: the central zone that rotates at the same velocity as the impeller, and the rest of the tank where the velocity is dumped as the liquid approaches the walls. The tangential velocities are given by:

$$u_t = \begin{cases} u_{t1}(r/r_1) & r \leqq r_1 \\ u_{t1}(r_1/r) & r > r_1 \end{cases} \tag{6}$$

and r_1 is a characteristic radius, the parameter of the model, for which

$$u_{t1} = 2\pi r_1 N \tag{7}$$

Calculation of the shear rate deriving Eq. (6) respect to the radius r leads to a value of zero in the core, where the liquid moves as a solid rod and $4\pi N$ in the outer region. This gives $k = 4\pi$ in Eq. (3) which is very close to the value $k = 12$ generally accepted for flat blade turbines.

Other criteria have also been applied, in order to characterize the shear in a tank and be able to design or scale-up a process.

Sinskey et al. [25] defined an integrated shear factor (ISF), as

$$ISF = 2\pi ND_i/D - D_i) \tag{8}$$

as a measure of the strength of the shear field between impeller and wall. Hu [26] tested the criterion for different sizes of similar vessels and impellers, and found that the relative extent of growth of mammalian cells attached to microcarriers was satisfactorily represented by a single line when plotted against the ISF as defined in Eq. (8), for all his vessels and impellers, suggesting that the ISF, even if appearing as arbitrarily defined and lacking of basic theory, could be used for scaling up within the range of sizes used (at least 0.250 to 21). A critical ISF can be defined as the limit that will assure that cell growth is not affected.

Crougham et al. [27] made an attempt to define a more fundamental method for characterization of the shear field. They adopted for their spinner flasks the following tangential velocities from Nagata [28]

$$u_t = \begin{cases} 2\pi Nr & 0 \leqq r \leqq r_c \\ 2\pi N(r_c/r)^{0.8} & r_c \leqq r \leqq D/2 \end{cases} \tag{9}$$

and the parameter r_c is given as a function of the Reynolds number

$$r_c/r_i = Re\,(1000 + 1.6\,Re) \tag{10}$$

Calculating the average shear rate on the bioreactor, they arrive to the expression:

$$\gamma_{av} = \frac{112.8 N r_i^{1.8}[(D/2)^{0.2} - r_i^{0.2}]\,(r_c/r_i)^{1.8}}{(D/2)^2 - (r_i)^2} \tag{11}$$

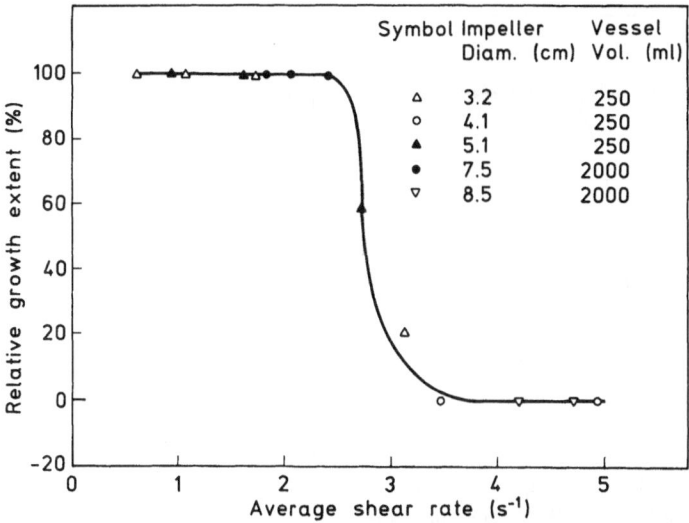

Symbol	Impeller Diam. (cm)	Vessel Vol. (ml)
△	3.2	250
○	4.1	250
▲	5.1	250
●	7.5	2000
▽	8.5	2000

Fig. 9. Relative growth of FSU cells and chicken embryo fibroblasts vs, the average shear rate γ_{av} [27]

Calculating the γ_{ar} for experiments on FS4 cells (2 sizes of bioreactor) and for chicken embryo fibroblasts (2 sizes of bioreactor), they found that all data for each cell line could be presented as a single line on a plot of relative cell growth versus γ_{av}. Figure 9 shows these results and suggests that a critical γ_{ar} can be defined for this type of bioreactors for each cell line, in order to assure that shear will not affect cell growth.

An alternative analysis by the same authors attempts the interpretation of the effect of shear on cell viability in a turbulent field using Kolgomorov's model of isotropic equilibrium.

A length scale and a velocity scale are defined as:

$$L = (v^3/P_m)^{1/4} \tag{12}$$

$$v = (v/P_m)^{1/4} \tag{13}$$

where P_m is the local power dissipation per unit mass, and can be obtained from correlations in the literature [29].

They succeeded to represent their cell growth data as a function of the length scale, Fig. 10. The figure suggests that when the length scale L goes down below a certain value, the cells are affected by turbulence. The critical L was found to be in this case near 100 μm, which is close to the diameter of the microcarriers used. It would be expected that the same turbulence level should not affect free cells in suspension, since energy would be dissipated at a scale that exceeds their size by an order of magnitude or more. In general, these results indicate that a scale-up in geometrically similar vessels should not be detrimental to the cells if the specific power input is maintained constant. These results also explain why there is a poor correlation between Reynolds number and shear damage. The Re is a rough measure of the macro-scale

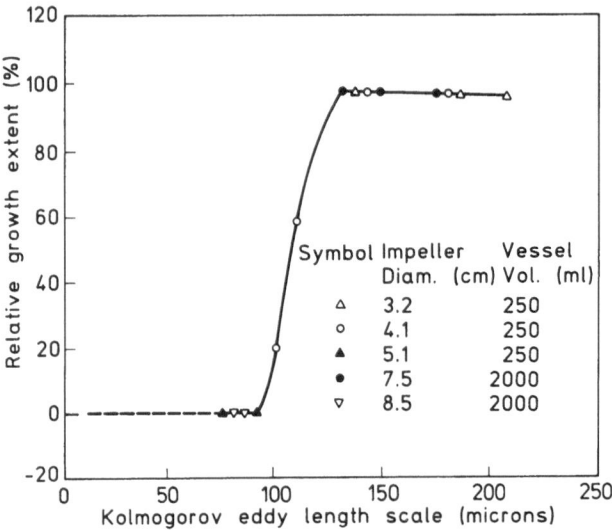

Fig. 10. Relative cell growth versus Kolgomorof's length scale L [27]

turbulence, turbulence in the form of eddies of the order of magnitude of the impeller blades. But most damage is caused by the microscale eddies.

It must however be kept in mind that the above experiments were all performed in relatively small bioreactors. It is to be expected that in a larger, industrial size plant, the required volume of a vessel for a biological reaction will be of such size that surface aeration will not be enough and the model will probably not be extrapolated to a sparged vessel. Nevertheless, the introduction of the microscale turbulence as a criteria of the detrimental limit that leads to cell damage is an important conceptual step by itself.

2.2 Shear in a Pneumatically Agitated Vessel

The first attempt to define and calculate a characteristic shear rate for a bubble column was done by Nishikawa et al. [30]. The calculation was done indirectly, through the heat transfer coefficient, as follows. A correlation was found for the heat transfer coefficient h vs. viscosity in Newtonian liquids. Then, the heat transfer coefficient for a non-Newtonian liquid was experimentally obtained. In a correlation for Newtonian liquids, the hypothetical viscosity μ_a giving the same h was found, as shown in Fig. 11. This apparent viscosity was then used on the flow curve of the non-Newtonian liquid (τ vs. γ) to obtain a shear rate that they defined as the average shear rate in the bubble column, γ_{av}. This was correlated with the superficial gas velocity, giving:

$$\gamma_{av} = 5000 \, J_G \tag{14}$$

Nakano and Yoshida [31] used this equation in order to calculate the apparent viscosity of a non-Newtonian liquid, and used this apparent viscosity to successfully

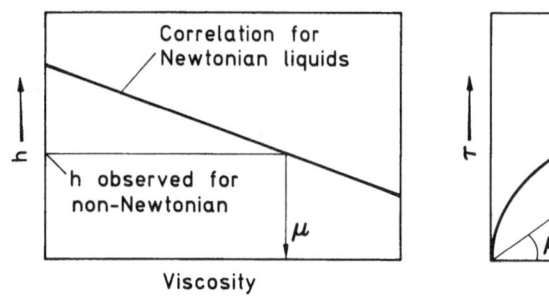

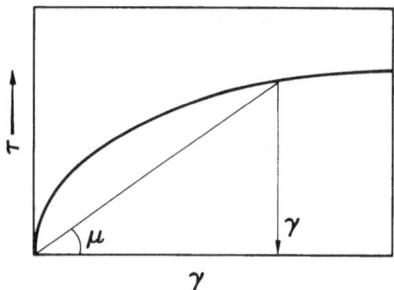

Fig. 11. Calculation of γ_{av} for non Newtonian liquids [30]

correlate their data of mass transfer rate. Such success was not obvious, since the heat transfer coefficient refers to a phenomenon occurring between the liquid and the wall of the column, and therefore could be expected to represent a mean value for the wall shear stress. The mass transfer, on the other hand, takes place at the gas-liquid interface around the bubbles, which are mainly far from the wall. Nevertheless, Nakano and Yoshida's success encouraged others to calculate γ_a in bubble columns and even in air lift reactors using the same correlation [30, 31].

Stein [24] proposed a semi-theoretical model that takes into consideration that energy may enter the system both with the gas and due to a mechanical stirrer, and derived:

$$\gamma_{av} \propto (J_G \varrho_L g + P/V)^{0.335} \, J_G^{0.245} \, \mu_{ap}^{-0.115} \tag{15}$$

If no mechanical agitation is used, $P/V = 0$ and the exponent of the superficial velocity is 0.58. Thus, this expression predicts an entirely different dependency of γ_{ap} on J_G. In addition, μ_{ap} has also some weight. This confirms the observation that by adding carboxymethyl cellulose to a cell culture in order to enhance the viscosity, the damaging effect of mechanical agitation can be reduced [27]. Up to now, no devices for experimental evaluation of γ_{av} in mass transfer operations have been reported, and therefore the clarification of actual dependence of γ on J_G is pending.

An alternative to the above approach is the development of mathematical models to describe the flow field in a pneumatic reactor. Ueyama and Miyauchi [34] solved the equation of motion for the two-phase flow within a bubble column by considering time averages of the Navier-Stokes equation and later extended the formulation to the recirculation flow regime [35].

Yang et al. [36] developed a model based on the concepts proposed by Ueyama and Miyauchi [34]. They arrived to the differential equation:

$$-\frac{1}{r} \frac{d}{dr} \left(\mu_t r \frac{du_L}{dr} \right) = \frac{4}{D\varrho_l} \tau_w - (\varepsilon_{av} - \varepsilon) g \tag{16}$$

where μ_t is the turbulent viscosity. Here it is assumed that the hold-up ε is a function of the radial position

$$\varepsilon = \varepsilon_{av} \left(\frac{b+2}{b} \right) (1 - \varphi^b) \qquad 1.8 < b < 2.3 \tag{17}$$

where φ is a dimensionless radius

$$\varphi = 2r/D \tag{18}$$

and therefore $\varphi = 0$ at the axis and $\varphi = 1$ at the wall.

In the integration of Eq. (16), Yang et al. [36] added the following condition to the boundary conditions of symmetry and a maximum in w_L at the axis:

$$u_L = 0 \qquad \text{at} \qquad \varphi = Q \tag{19}$$

Equation (19) implies that there is an inversion of the liquid flow at a certain point $\dot{Q}$, which is defined by Eq. (19), and is in fact the parameter of the model.

The solution given by the authors is:

$$u_L = u_{L\,max} \left[1 - Q^{-2}\varphi^{-2} + \frac{2 - Q^{-2}}{Q^b - 4/(b+4)} (\varphi^{(b+2)} - Q^b\varphi^2) \right] + J_G \tag{20}$$

The shear rate can be obtained by derivation of Eq. (20) respect to φ, giving:

$$\gamma_d = \frac{d(u_L/u_{L\,max})}{d\varphi} = 2\varphi Q^{-2} + K[\varphi^{(b+1)}(b+2) - 2\varphi Q^b]$$

where γ_d is a dimensionless shear rate: $\gamma_d = \gamma \dfrac{D}{2w_{Lmax}}$ $\qquad\qquad$ (21)

The average shear rate in the column can now be obtained as:

$$\gamma_{av} = \frac{4}{D^2} \int\limits_0^{D/2} |\gamma(r)|\, r\, dr = \int\limits_0^1 |\gamma_d(\Phi)|\, \Phi\, d\Phi \tag{22}$$

$$\gamma_{av} = -\frac{2}{3} Q^{-2} - \frac{2}{3} KQ^b + K\left(\frac{b+2}{b+3}\right) \tag{23}$$

The absolute value of γ must be taken into account, as indicated in Eq. (22). Therefore. Eq. (23) is applicable only over regions where γ_{av} does not change sign. Each such region should be considered separately, and an overall average calculated for all the range. Fig. 12 shows profiles of shear rates in a bubble column calculated from Eq. (21). It can be seen that in a certain range of the parameters of the model, γ_d does change sign, and therefore special care should be taken when using Eq. (23).

While the model proposed by Yang et al. [36] has the advantage of the simplicity of the resulting equations, it has the disadvantage that the main variable of the system, the air flow rate, does not appear explicitly. It is represented instead by Q, the parameter of the model.

Other models for bubble column fluid dynamics give more weight to the interaction between bubbles and liquid. Molerus and Kurtin [37] presented a model based on integral flow parameters, taking into account the recirculation of small bubbles

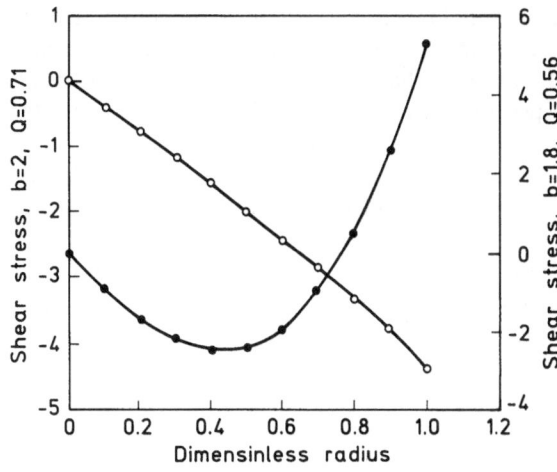

Fig. 12. Profiles of shear rates in a bubble column, calculated from the model by Yang et al. [36]. ● Dimensionless shear rate for b = 2, Q — 0.71
○; Dimensionless shear rate for b = 1.8, Q — 0.56

with the liquid in the recirculation regime. Joshi and Sharma [38] presented a model based on the formation of annular eddies in bubble columns, which can be visually detected. Zehner [39] extended this model for annular to cylindrical eddies. These models are realistically based on the actual flow configuration in the reactor. The maximum liquid velocity can be predicted, but the obtention of the complete flow field is not straightforward, and therefore derivation of mean shear rates is difficult.

In general, the task of evaluating shear rates profiles in pneumatically agitated vessels still deserves much attention and efforts from the scientific community.

3 Shear Effects on Microorganisms

Shear can obviously be a disrupting factor in microorganism culture. But it is also related to heat and mass transfer, and in its absence, all transfer would have to be by pure molecular mechanisms, thus very slow. It follows that an optimum shear level exists for a given culture. It will depend on the resistance of the cell to mechanical forces and the extent of deformation it can stand undamaged, on the levels of requirements of nutrients and on the potentially detrimental effects of metabolites in each organism.

But in addition to the above mentioned phenomena, changes in the morphology of microorganisms, which are associated with the shear field in the environment, have been observed in many cases. The relationship between such morphological changes and rates of growth and metabolite production has still not been properly understood, although it may be of great importance in the design and scale-up of bioreactors.

The recognition of the effects of mixing on morphology is not new, although many times forgotten. One of the first testimonies was given by Camposano et al. [40] on the effects of agitation on the morphology of *Aspergillus flavius* and the production of kogic acid in submerged cultures. It was found that at too high a rate of agitation, the mycelium formed is short and strongly branched. It was also found that, hand in hand with the morphological changes, mainly starch rather than kojic acid was formed.

The response of a filamentous fungus to a change in agitation conditions depends also on pH, temperature, and especially, osmotic potential, which influences the hyphal turgor pressure. Furthermore, it has been shown by Pitt and Bull [41] that even if changes in agitation rate produce only small changes in the macromolecular composition of filamentous fungus, the number of growing tips is sensibly affected by such changes.

Metz [42] found that the length of the main hyphae of *P. chrisogenum* in a continuous culture changed with the stirrer speed. It was approximately 200×10^{-6} m at 1000 rpm, and stabilized at 150×10^{-6} m at 2100 rpm. Decreasing the agitation speed to the range 450–800 rpm would then double the length of the main hyphae (Fig. 13).

Such morphological changes suggest that agitation conditions influence also the kinetics of biomass growth and metabolites production. The work of Konig et al. [43] shows that while the rate of biomass growth decreases constantly as agitation decreases, the rate of penicillin production and substrate consumption increases to a maximum and then decreases again. Similar observations on citric acid production by *Aspergillus niger* were reported by Ujcova et al. [44]. This can be understood as the direct effects of mechanical forces acting on the hyphae, and the transfer of substrates and products which increases with the increase of the shear rate in the solution.

These effects of mechanical forces are not exclusive of filamentous fungae, but rather seem to apply to any elongated form. The experiments of Bronnenmeier and Märkl [45, 46] show the effect of shear stress on the alga *Spirulina platensis*. Increasing the levels of shear stress lead first to changes in the structure of the microorganisms and then to break up.

An alternative way for the evaluation of shear damage was proposed by Tanaka et al. [47]. The light extension at 260 nm in the cell-free suspension was measured and

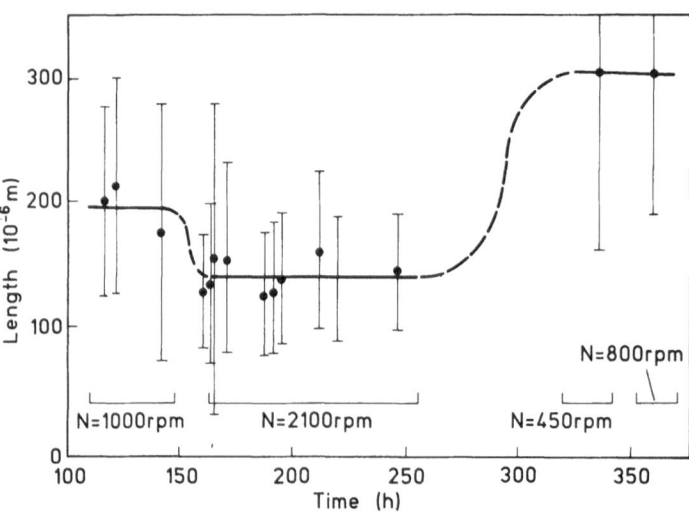

Fig. 13. Variation in the length of the main hyphae during a continuous experiment at varying stirrer speeds, with *Penicillin chrysogenum* [42]

considered to be proportional to the amount of nucleotides leaked into the media. Such measure would also take into account damages to the cells that do not lead to the rupture of the hyphae but do injure the cell wall. Their results showed an increasing amount of nucleotides in solution as the agitator speed increased.

Reuss [48] showed that the mechanical damage to *Rhizopus nigricans* in a stirred bioreactor can be represented as a function of the ratio of power input to flow rate. This model is based on the analogy with processes of mechanical disintegration. As mentioned before, in the present review we are interested in sublethal effects rather than in the recovery of intercellular products. However, it is possible that this approach may be extended to specify limits of shear on given cultures.

The effects of agitation rate on the morphology of *E. coli* have also been reported (Wase and Ratewate [49]). The mean cellular volume increased strongly with the increase in stirrer speed or stirrer dimensions. The same effect was found in different bioreactor configurations and under both carbon-limited and nitrogen-limited culture conditions. These observations were repeated by Wase and Patel [50] on cultures of *B. cereus*, *Staph. epidermidis* and *Sacch. cerevisiae*.

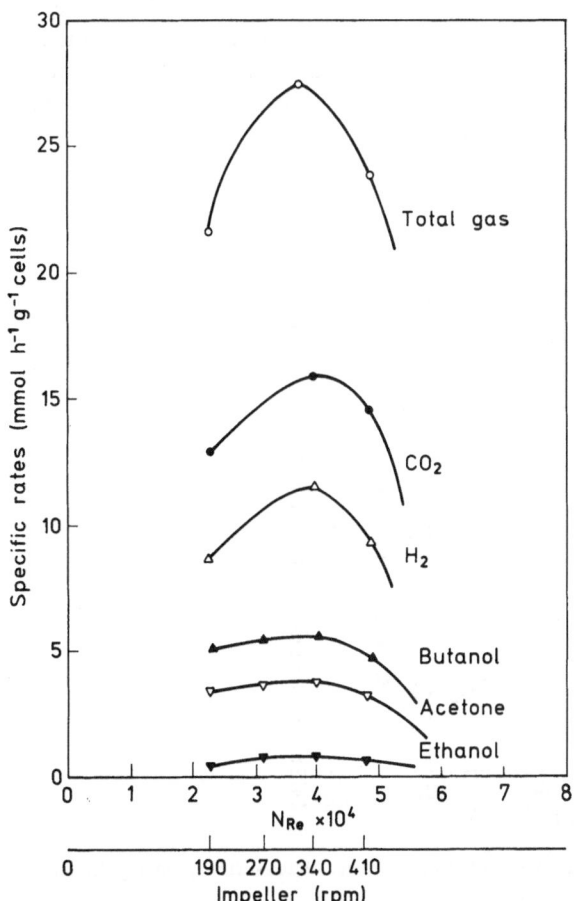

Fig. 14. Variations of specific production rates of solvents and gases with variation in impeller rotational speed in an acetone-butanol process of *Clostridium acetobutyricum* [51]

But the observations on the effects of agitation conditions are not limited to morphological changes. Yerushalmi and Volesky [51] studied the effects of the speed of agitation on the rates of formation of products in the acetone-butanol process by *Clostridium acetobutyricum*. They reported that the concentrations of carbon dioxide, hydrogen, butanol, acetone and ethanol increased as the impeller speed in their reactor increased, reaching a maximum near 350 rpm and decreased with further increase of the agitation rate (Fig. 14). These data, interesting as they are, cannot be taken as an absolute proof of direct shear effects on the microbial metabolism, since mass transfer phenomena may also have an influence, especially through the disposal of inhibiting metabolites from the surroundings of the cell. Such is the case presented by Funashi et al. [52], studying the production of Xantan gum. They reported a strong increase of specific production rate of Xantan gum as the agitation rate increased. This increase correlated fairly well with the tip velocity of the stirrer ND. However, a series of independent experiments, where the shear stress remained constant, showed a very good correlation of specific production rate and glucose concentration, suggesting that the transfer of substrate to the cells through the very viscous medium surrounding the cell may be the reason for the beneficial effect of the increase in the agitation rate.

Another example, where the change in metabolite production is clearly seen to be associated with the stirred speed and the morphological changes, is presented by McNeil and Kristansen [53]. They cultivated *Aureobacidium pullulans*, a dimorfic fungus that can assume both filamentous or yeast-like form to produce the extra cellular polysaccharide pullulan. It is known that the yeast-like cells are the ones that

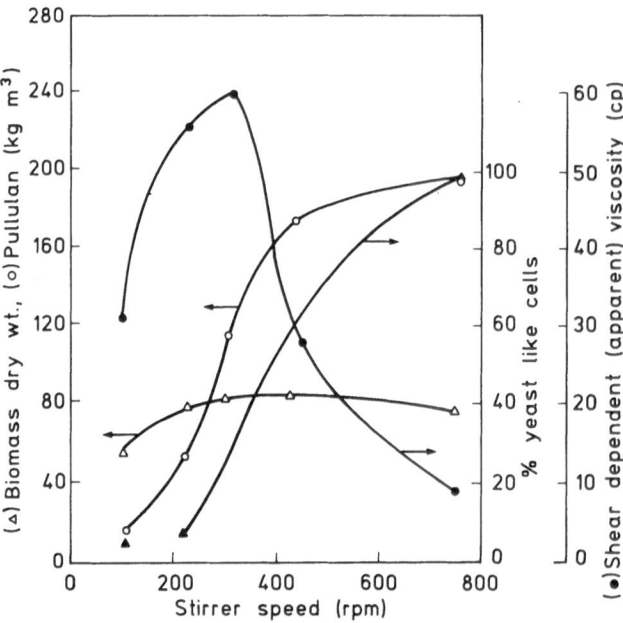

Fig. 15. Biomass, pullulan, yeast-like cells and apparent viscosity versus stirrer speed in pullulan production by *Aureobasidium pullulans* [53]

produce the pullulan. Figure 15 shows that as the stirred speed increases above 200 rpm, the culture switches increasingly to the yeast-like form and the pullulan concentration increases accordingly, while the apparent viscosity decreases due to the morphological changes.

Silva et al. [54] showed that *Dunaliella* cultures were increasingly sensitive to specific bubbling rates. Addition of carboxymethyl cellulose and agar to the cultures protected the microorganisms from the effects of shear stress. It will be shown later that the effects of bubbles has also been considered and studied for the case of mammalian cells in suspension.

It can be summarized from the above that there is strong evidence in the literature of the influence of the shear on the morphology of suspended microorganisms. The dependence of product formation rates on the shear in the medium is strongly suggested, but it is difficult to establish assertively due to the possibility of simultaneous mass transfer effects.

4 Shear Effects on Mammalian Cells

The potential of the mammalian cell cultures and recombinant DNA technology is impressive, covering viral vaccines, immunoregulators, monoclonal antibodies, polypeptide growth factors, enzymes for human therapy, hormones, tumor specific agents, etc. The surge of this new technology was the main reason for the focusing of research and development of new bioreactors on the issue of shear related damage to suspended cells. This is because in mammalian cells the rigid cell wall that protects and separates microorganisms from the medium, is missing. In addition to this, the larger size and doubling time (one order of magnitude in both) contribute to the vulnerability of mammalian cells.

Here, as in the case of microbial cultures, the dependence of morphology on the fluid dynamics can be clearly shown [55]. It was found clear evidence of changes in shape and orientation of the cells in the response of monolayers of vascular endothelial cells to a constant shear rate on a specially designed cone and plate apparatus. At higher shear rates, the cells adopt an elongated form and align themselves in the direction of the flow. If the conclusions of Folkman and Moscona [56], stating that the shape is tightly coupled to DNA synthesis and growth, are taken into account, the direct influence of shear rate on the growth and metabolic activity of the cells emerges. This is confirmed by published studies on aortic histamine synthesis. Hollis and Ferrone [57] showed that the histamine synthesis in a rabbit thoracic aorta increased sharply as the shear rate increased, (Fig. 16) and that over 50 % of this increase occurred in the endothelial monolayer. De Forrest and Hollis [58] confirmed these findings and proposed that the shear — stimulated local rate of aortic histamine formation may influence the wall permeability to blood-borne macromolecules such as albumin.

Stathopoulos and Hellums [59] designed a system of rectangular section, with embryonic kidney cells attached to the walls. The shear on the wall could be calculated (neglecting end effects). They worked in the laminar range, and the wall shear forces varied from 0.26 to 5.4 N m^{-2}. The production rate of urokinase remained constant at low values of the shear rate, but further increases in flow rates lead to an increase,

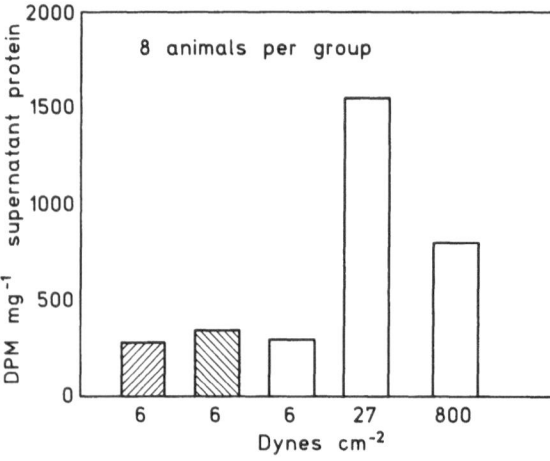

Fig. 16. Rabit thoracic aortic histamine-forming capacity (HFC) following exposure to 1 h shearing stress [57]

a maximum and a slow decrease of urokinase production. Changes in morphology were also observed in such range.

These research works were focused on the generation of arteriosclerosis, which is known to be a focal disease, starting with points of maximum shear in the cardiovascular system. The results indicating the link of the hemodynamic state and wall permeability transcend this field, however, and may be very important in industrial processes, where mass transfer rates of substrates and products through cell membranes may be the controlling step.

Frangos et al. [60] developed a flow apparatus where the response of endotelial layers of human umbilical vein to a wide range of shear stress was studied. The device allowed both steady and pulsatil flow under well defined and controlled conditions. They found that the onset of flow produced a sharp increase in prostacyclin production rate, which then decayed to a lower steady state value. Their results are a confirmation that, in certain ranges, shear rates may increase metabolite production.

Petersen et al. [61] subjected CRL-8018 hybridoma cells to shear in a specially designed bob-and-cup viscometer for short times, after having been cultivated in a bioreactor. They observed that the cell death in the viscometer showed similar trends as cell death caused by excessive agitation in spinner flasks, and concluded that viscose shear is the main cause of cell damage. They also observed that cells cultured with low levels of fluid stress were more sensitive to shear, suggesting a strong adaptation effect.

Handa et al. [62] studied the detrimental effects of sparger aeration on suspended mammalian cell cultures. They managed to obtain a video record of cells in the region of bubble disengagement at the medium surface in a bubble column. In nonfoaming liquids, cell damage seems to be associated with bubble bursting and consequent velocity fluctuation in the liquid film. In a foaming medium, damage is reported to be due to physical shearing effect in the draining of liquid films around the bubbles.

The authors found that when a stable foam is present, the cells do not penetrate the foam layer, and are therefore protected against the damaging effects of bursting bubbles and film drainage. This work seems to indicate therefore that the cell damage in bubble bioreactors is not associated directly with the gas-liquid interface, but rather with liquid velocity fluctuations, especially in the region of bubble disengagement.

From the above-mentioned publications, it becomes clear that shear may have detrimental effects on mammalian cells, but in certain regions changes occur both in morphology and in metabolic rates, before these effects become lethal. In particular, the production of metabolites and the permeability of cells to macromolecules have been studied. Since these studies were performed in the framework of arteriosclerosis research, the data correspond to cells adhering to a wall or forming part of it. This is enough, however, to suggest that also in suspended cultures, this type of effect is bound to be found.

5 Shear Effects on Plant Cells

Plant cells are much larger than microorganisms, usually in an order of magnitude. In addition, their cellulose wall is rigid. It follows that an analysis in terms of Kolgomorov's model of isotropic turbulence will indicate that serious damages are to be expected at relatively large values of the length scale L given in Eq. (12) [26]. Indeed, it has been observed that plant cells are shear sensitive, and difficulties have been found in their cultivation in stirred tank bioreactors [63, 64]. This is especially true when large scale systems are considered. But while high agitation rates may be detrimental to cell growth, low agitation will increase the amount and size of cell clumps, originated both by daughter cells that stay together after division, and by the stickiness of the polysaccharides that the cells excrete, especially at the end of the growth phase. An optimal shear rate must be found for each culture, by maneuvering between these two extremes. This stresses the relevance of the establishment of critical values for shear in cell plant cultures, and the selection of shear resistant cells [65].

Scragg et al. [66] studied the effect of agitation rate in a stirred bioreactor on cell suspensions of Catharanthus roseus and Helianthus annuus. This was done by exposing the cell cultures to fixed impeller speeds in a 3-liter stirred tank bioreactor for various periods of time. Samples were taken at intervals, and inoculated into a fresh medium in a 250 ml flask. Viability was estimated by comparing the initial mass of the sample to the mass after 14 d of incubation. The authors found that several of the cell lines studied were able to grow in stirred tank bioreactors at low agitation rates (150–200 rpm) after being held for several hours at 1000 rpm. They further confirmed the very important fact that shear tolerance is not a fixed characteristic of the cell line, but in some cases can be developed during culture under shear conditions.

These experiments, enlightening as they are, have the shortcoming that they just check the effect of a relatively short time of high shear application, and thereafter the culture is transferred into a much gentler environment, where the capacity of growth is observed. The scope of the present work however, is to inspect the effects of shear for long periods of time, namely several doubling times, which would allow the adaptation processes to occur.

The approach of considering the bubbles as the main responsible for cell damage, similar to that of Handa et al. [72], hals also been used as criteria for design of fito-reactors. This is usually done by defining two separate zones, one enclosing the plant cells and the other for gas exchange by bubbling.

A different and interesting approach to the problem of evaluating the shear acting on suspended cells is due to Tanaka [63]. It is based on the measurement of the mass transfer coefficient from the liquid to the suspended solid, which is a good measure of the shear rate at the liquid-solid interface. The use of this coefficient as a measure of agitation efficiency had already been proposed by Hixson and Crowell [67]. Tanaka measured the rate of dissolution of β naphthol particles suspended in the liquid, and showed that the cell mass concentration of a 15-d old culture, both in shake flasks and Erlenmeyer flasks, were a decreasing function of the liquid-solid mass transfer coefficient K, as shown in Fig. 17. The figure indicates that under the author's experimental conditions, the cell growth was mainly limited by hydrodynamic stress. It is

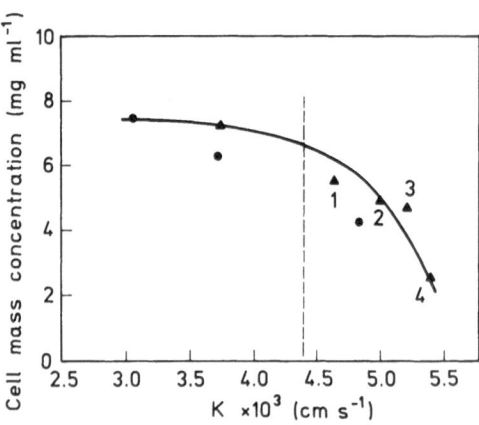

Fig. 17. Relation between the cell mass concentration of a 15-d old culture of *Cudrania tricuspidata* and the mass transfer coefficient of β-naphtol-water system, K, at 30 °C [63]. Numbers in the figure indicate the number of baffles plates in Erlenmeyer flasks. (○) shake flasks, (△) Erlenmeyer flasks

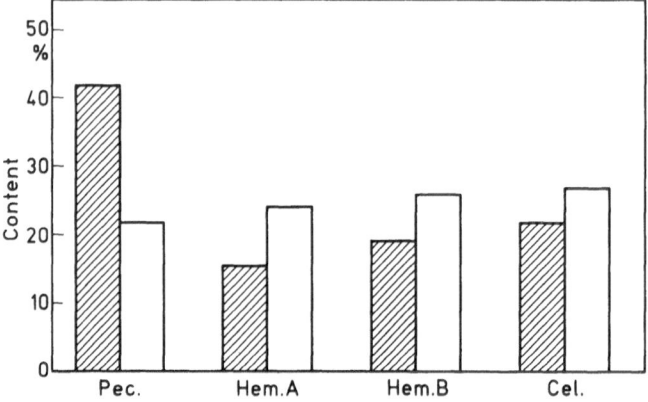

Fig. 18. Amount of cell wall components of cells of *Catharantus roseus* using flasks. Pec = pectin, HemA = Hemicellulose A, Hem B = Hemicellulose B, Cell = Cellulose, ▨: without baffle plates; ☐: with baffle plates [68]

to be expected that if the graph is extended to the lower range of K, the cell mass concentration would decrease again, as a zone of mass transfer limitation is reached, establishing therefore an optimal range of operation from the point of view of the hydrodynamic stress. This reinforces the thesis that shear must be considered as one of the variables in the kinetic equation of suspended cells.

Tanaka et al. [68] studied the effects of stress generated by aeration and agitation in suspended cultures of *Catharantus roseus* cells, and found that physical hydrodynamic stress was closely related to growth rate, maximum cell mass and size distribution of cell aggregates. But the main finding, from the point of view of the present review, was that the cell wall components changed also as a function of the hydrodynamics. Figure 18 shows how the amounts of pectin, cellulose and hemicellulose changed when the cells were cultivated, under the same conditions otherwise, in flasks with an without baffles. It can be expected that changes in the wall composition affect the kinetics of growth and metabolic permeation. This is parallel to the findings of De Forest and Hollis [58] on the permeability of endothelial cells, mentioned in Sect. 4.

6 Methods for Evaluation of Shear Effects

The evidence shown in the sections above on the effects of shear on both morphology and kinetics in microorganisms, mammalian and plant cells suggest that hydrodynamics should be considered as one of the variables in the kinetic formulations used for the design of bioreactors. In order to obtain a clear quantitative relationship between shear and growth or production rates, kinetic studies must be performed in systems with defined and known flow characteristics. Such were the experimental systems used by Hollis and Ferrone [57] and Stathopoulos and Hellums [59]. In both cases the geometry of the system (a cylinder and a channel of constant rectangular section) allowed the calculation of the shear stress on the walls during laminar flow. Since the studies were focused on cells attached to the surface, as explained above, the effects observed could be related quantitatively to the shear rates acting on the cells. But this quantitative information can be taken only as an indication of general trends from the point of view of the present paper, which handles with suspended cell cultures.

A system with a known and homogeneous shear field is required, in order to correlate morphological or physiological changes to a certain shear rate or shear stress. The main geometries which have been used are the cone-and-plate and the concentric calinders, both geometries being used for rheological measurements. The cone-and-plate devices assure a constant value of the shear rate in all the volume [69]. In concentric cylinders the shear rate changes in the gap as a function of the radius. But the gap must be kept small in order to assure a laminar flow. The condition for a laminar flow without the formation of "Taylor vortices", internal eddies which appear for a rotating inner cylinder, is satisfied when the Taylor number is:

$$Ta = \frac{\Omega \delta R_1}{\nu} \sqrt{\frac{2\delta}{R_1 + R_2}} > 31.4 \qquad (24)$$

where δ is the gap between cylinders, v the kinematic viscosity, R_1 and R_2 the radius of the internal and external cylinder respectively, and Ω is the revolution speed.

As said above, and expressed in Eq. (24), a small gap assures laminar flow. It follows therefore that the shear rates in the gap will be practically constant, since the radius is much larger than δ.

Midler and Finn [70] did a pioneering work in a concentric cylinders device. Their objective was the evaluation of shear effects in the disruption of microorganisms. They used a protozoa (*Tetrahymena pyriformis w*), which was considered to be considerably more sensitive to shear than plant cells. They used in parallel stirred vessels and their Couette-type device (Fig. 19). They worked at constant angular velocity and with a small gap between cylinders, which varied from 0.63 to 1.2 mm. Since it was found that the protozoa are quite resistant to shear, they sensibilized them by putting them in a 5 % salt solution; the animals were dead but retained the shape and integrity, and were destroyed when mechanically disturbed. It was found, in the stirred vessels, that the variable correlating satisfactorily the disruption rate was the tip velocity. In the constant shear device, damage was observed only above a critical shear rate of 1200 s^{-1}.

Working with different gaps and viscosities, they concluded that the proper variable to correlate the death rate was the shear rate.

Bronnenmeier and Märkl [45] designed a different type of device, considering that the destructive mechanisms beyond hydrodynamic stress in stirred vessels are two: 1) rapid pressure change over the blade, accompanied by acceleration of the microorganisms; and 2) high shear stresses at the turbine tips. They adopted a free jet as a model with a defined flow in which both pressure changes and shear stress are present (Fig. 20). Their experiments consisted of the application of hydrodynamic stress via

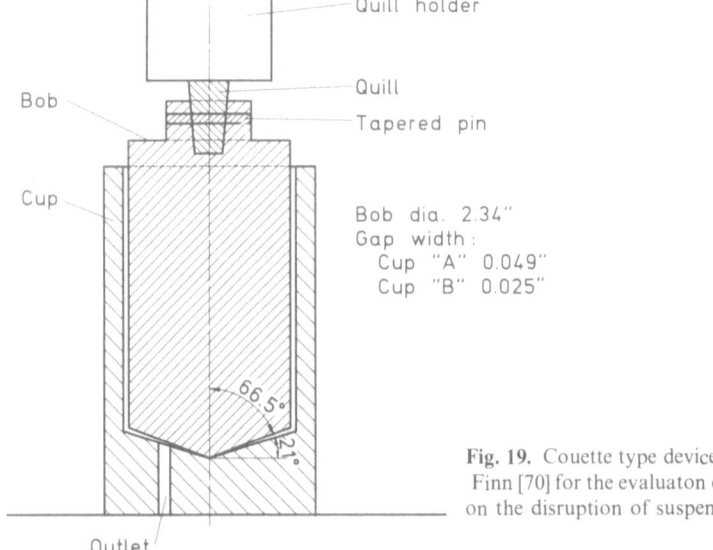

Fig. 19. Couette type device used by Midler and Finn [70] for the evaluaton of the effects of shear on the disruption of suspended microorganisms

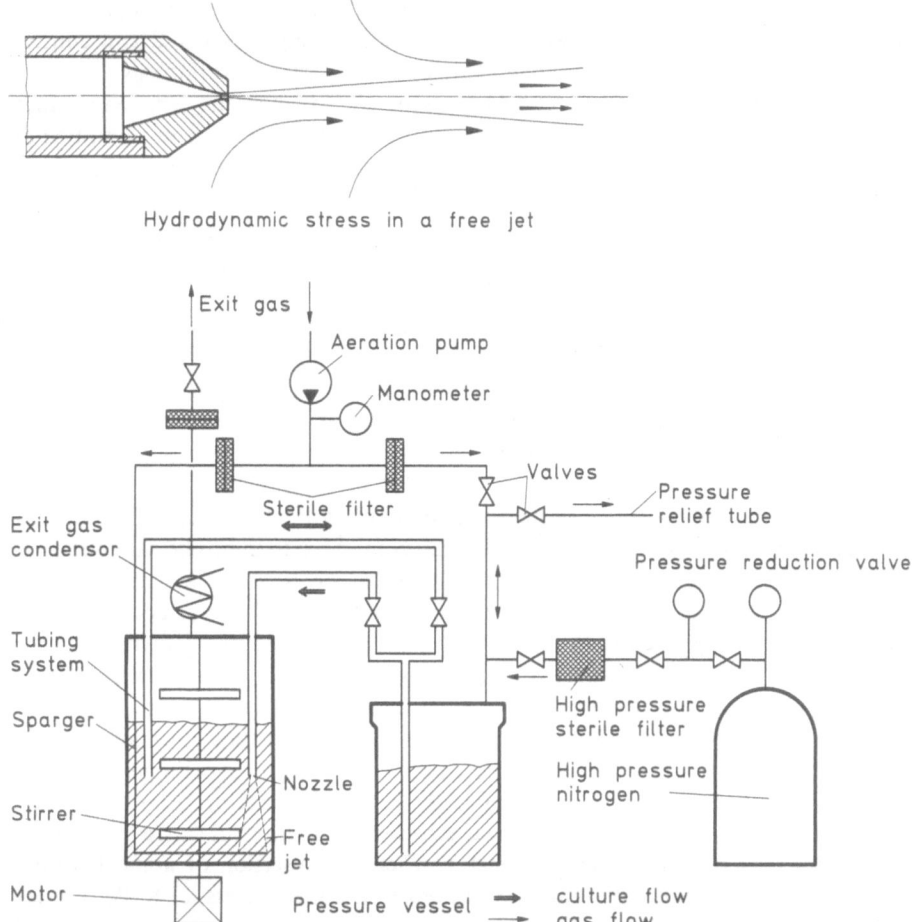

Hydrodynamic stress in a free jet

Fig. 20. Free jet test apparatus for hydrodynamic tests on suspended microörganisms [45]

the free jet apparatus, and observations on the culture evolution thereafter. They found it difficult, however, to compare their experiments on three green algae and two cyanobacteria with cultures grown in standard stirred vessels, obviously because while the stress is continuous in the stirring experiments, it lasted only for a short time in the free jet apparatus.

Smith et al. [71] used a standard viscometer in order to study the shear sensitivity of mouse hybridoma cells producing antibodies to surface antigen of *Brucella abortus*. Cultures were transferred from a continuing spinner flask to the viscometer, which was first summarily sterilized with 70% alcohol, and samples were taken hourly during runs that lasted 15 h. Experimental results showed that while at a shear rate of 424 s^{-1} the culture behaviour did not deviate from the behavior of the control, at 869 s^{-1} there was a sharp difference in live cell count with time; the cell division seemed to be inhibited, and the cell count decreased, in contrast with a continuous increase in the control culture: this is coherent with lower cell viability and a sharp increase in the concentration of the intracellular enzyme lactate dehydrogenase, an indicator of cell disruption. However, SEM studies of highly sheared and control hybridoma cells did not reveal distinct changes in cell surface morphology.

Schürch et al. [72] designed a sterilizable viscometer in which they studied laminar shear stress effects on hybridoma mass cell cultures producing monoclonal antibodies against mithochondrial creatine kinase. The cells were subjected to shear in the viscometer for short intervals of time (2 to 10 min) and afterwards their performance was studied in spinner flasks, following viable cell counts, death rate, pH, antibody concentration, lactate and glucose concentrations. Their results show identical doubling time, glucose uptake, and lactate synthesis for sheared and unsheared cultures, and these results are parallel to thos obtained by Scragg et al. [66] for plant cells. In both cases, cultures were put under acute stress conditions for short periods of time and then their behaviour studied under milder conditions. The only clear effect reported by Schürch et al. is an increase in cell damage. Fig. 21 shows that the percentage of cell damage increases from 1 to 3 % when the applied shear stress is increased up to 16 N m^{-2}.

Chittur et al. [73] subjected suspensions containing human T cells, B cells and monocytes to uniform shear stresses of 10 to 20 N m^{-2} for 10 min. The data obtained suggested that controlled exposure to sub-lethal shear stress results in alterations that can affect the proliferative response of the T cell population. This result contrast with the findings of Crougham and Wang [74] that for FS-4 cell cultures, growth is neither enhanced nor inhibited by hydrodynamic forces. It should be remembered however that the last authors refer to mildly agitated cultures.

Wudtke and Schügerl [75] studied the effects of hydrodynamic stress on ovarian cells of the butterfly *Spodoptera frugiperda*, used for growing insect pathogenic viruses. In order to cover the stress generated by both shear forces and pressure loads, they used a series of devices; a) Constant shear stress was generated in single and double gap viscometers. b) A modified viscometer, where the insertion of non-cylindrical rotors allowed to produce periodical variations in the shear stress. c) Roller bottles where cylindrical metal bars were introduced in order to generate short periodical pressure load on the cells attached to the inner surface of the bottles. d) A pressure-driven free jet apparatus. e) A bubble column with and without a paraffin oil layer

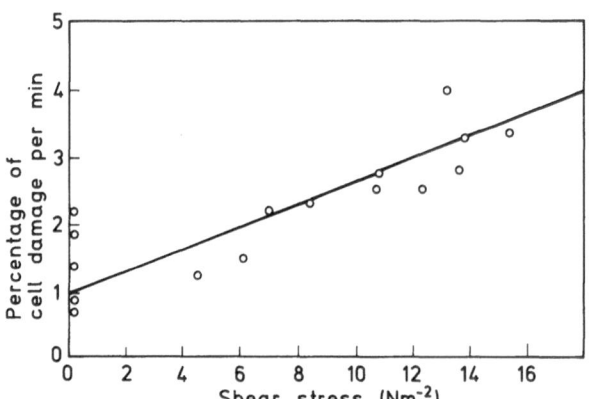

Fig. 21. Increase in hybridoma cell damage due to increasing shear stress applied in a laminar flow viscometer [72]

on the top in order to show the effects of bubble bursting. Viability of the cells was studied and their general conclusion was that those particular animal cells withstand a fair amount of stress. Their constant stress experiments were the longest, up to 24 h. The experiments in a double gap device, which allowed the generation of higher shear (up to 3680 l s^{-1}) produced an almost total loss of viability within 7 h. The most important point in this wok consists in the comparison of different shear generating devices, several of which were tested by other researchers as indicated above, with the same biological system. It is interesting to note that the bubble column was the deadliest of all. When a layer of paraffin oil was added in order to avoid the bursting of bubbles near the cells, the damage was much less acute. This is incomplete agreement with the findings of Handa et al. [62] on the effect of the rapid pressure fluctuation originated in the bursting of bubbles on mammalian cells suspended in the surroundings.

Merchuk et al. [76, 77] presented an improved design of a device for assessment of shear effects (DASHE) on suspended cell cultures. The device could be sterilyzed, and consisted of a combination of a coaxial cylinder plus cone and plate viscometers and enabled cultures of both eukaryotic and prokaryotic microorganisms to be grown under fully defined and controlled fluid dynamic characteristics for several generations. The DASHE consisted of two regions: one can be considered as a cone and plate system and the other as a coaxial system. The dimensions of the DASHE were so designed as to give the same shear rate in both the cone and plate plus coaxial cylinder regions. In the preliminary tests performed, a change in cell length was observed in *Escherichia coli*, and with *Penicillium chrysogenum* the septal length, hyphal diameter and branching frequency all changed as shear was increased. Figure 22 shows the shift in the mean cellular length in *E. coli* from the inoculum, grown in a shaken Erlenmeyer flask, to much larger cells grown in the constant shear device.

The main difference between this device and all those described before, is that while the later were intended to give a defined high shear environment to the cells for short

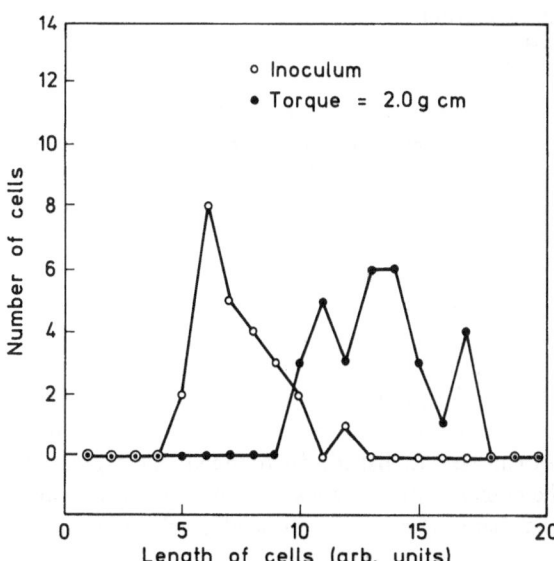

Fig. 22. Differences in the distribution of the length of *E. coli* grown in shaken Erlenmeyer flasks and the device for the assessment of shear effects (DASHE) proposed by Edwards et al. [76]

periods of time, in order to define a critical value for the hydrodynamic conditions before cell damage appears, this method allows the cultivation of a culture for long periods of time under defined shear conditions, simultaneously assuring sterility of operation and satisfactory gas exchange. The last condition is satisfied through a rigid gas permeable membrane which allowed sufficient oxygen transfer to insure aerobic conditions to cultures of both *Escherichia coli*, and *Penicillium chrysogenum*, at the relatively low biomass concentrations used. The device can therefore be used for prolonged times in order to evaluate the full adaptation of the culture to the shearing conditions. In the case of *Escherichia coli*, about 10 generations produced the enlarged cells refered to above. When inoculated again in the Erlenmeyer flasks, the original characteristics of the microorganisms were obtained.

It seems therefore that there are now available methods for the accurate evaluation of shear effects on growth rate and kinetic of product formation. Much work is still to be done till useful data will be available for many cultures.

7 Implications on Biorector Design

It has been shown above that many indications exist in the scientific literature suggesting the influence of shear on the kinetics of biomass and product formation. These indications have been brought from experiments with microorganisms, animal and plant cells.

The differences between these three types of cultures and even between different microorganisms, or different lines of cells, are so large that it does not seems reasonable to expect a common behavior.

But even if still far — off from a general theory that explains these effects, empirical information on the phenomena will be extremely important.

Experimental work has thus to be done on every species to clarify if in this particular case the hydrodynamics have an input into the kinetics of the process.

This knowledge is relevant for bioreactor design and scale-up, especially taking into account the considerations made at the beginning of this review on the need of a structural approach to the mathetical modelling and design of bioreactors.

Those structural models for bioreactors belong to the approach described by Young [78] as total environmental (TE) approach, based on the concept of an average microbial cell, environment and physiological state interaction. This concept implies that "a cell of a given genetic make-up, within a limited range of environmental conditions, has predictable and reproducible responses that can be qualitatively defined by a set of specific rates". The point that has been stressed in the present paper is that shear must be considered as one of the potential variables in the kinetic formulation of growth and product formation. Once this kinetic expression is available, a structured model of the bioreactor can be established, where the kinetic expression in each region must consider all the chemical and physical variables of the environment, and between the last ones the shear affecting the suspended cells and microorganisms.

8 List of Symbols

b   constant of hold-up distribution, Eq. (17), (—)

B   constant in Eq. (4) (—)

D   diameter of the reactor (m)

h   heat transfer coefficient ($J\ m^{-2}\ s^{-1}\ {}^\circ K^{-1}$)

ISF Integrated Shear Factor (—)

J_0 Superficial gas velocity ($m\ s^{-1}$)

K   liquid-solid mass transfer coefficient ($m\ s^{-1}$)

$k_L a$ volumetric gas-liquid mass transfer coefficient (s^{-1})

L   Kolgomorof's length scale, Eq. (12) (m)

n   given in Eq. (5) (—)

N   impeller speed (rpm)

P   power input to the bioreactor (W)

P_m local power dissipation per unit mass ($W\ kg^{-1}$)

Q   parameter in Eq. (4) (—)

r   radius of the reactor (m)

Re Reynolds number (—)

u   velocity ($m\ s^{-1}$)

V   volume of the bioreactor (m^3)

v   Kolgomorof's velocity scale, Eq. (13) ($m\ s^{-1}$)

v_x velocity in the x direction ($m\ s^{-1}$)

Greek letters

ε gas hold-up (—)

ν kinetic viscosity ($m^2\ s^{-1}$)

γ shear rate (s^{-1})

μ viscosity ($Pa\ s^{-1}$)

τ shear stress (Pa)

φ dimensionless radius, Eq. (18)

Subscript

a   apparent

av average

i   impeller

d   dimensionless

t   tangential

c   limit of vortex zone, Eq. (9)

v   per unit volume

w   wall

9 References

1. Atkinson B, Mavituna F (1983) Biochemical Engineering and Biotechnology Handbook. Macmillan, Suffolk
2. Bliem R, Katinger H (1988) Tibtech 6: 224
3. Oosterhuis NMG (1983) in: "Advances in Fermentation", Wheatland Journals Ltd., Hertsd., England, p 191
4. Oosterhuis NMG (1984) Ph. D. Thesis, Delft University of Technology
5. Kossen NWF, Oosterhuis NMG (1985) in: Biotechnology, Vol II, Chap. 24, Rehm NJ, Reed J (eds) VCH, Weinheim
6. Sweere APJ, Luyben KChAM, Kossen NWF (1987) Enzyme Microb. Technol. 9: 386
7. Johnston RE, Thring MW (1957) Pilot Plant, Models and Scale Up in Chemical Engineering, McGraw Hill, New York
8. Jordan DG (1968) "Chemical Process Development" Schoen HM and McKetta JJ (eds) Wiley, New York Vol. 6, part 1–2
9. Taguchi H, Imanaka T, Teramoto S, Takatsu M, Sato M (1986) J. Ferment Technol 46: 823
10. Takei H, Misusawa K, Yoshida F (1975) J Ferment Technol 53: 151
11. Roels JA (1983) Energetics and Kinetics in Biotechnology, Elsevier, Amsterdam
12. Bajpai RK, Reuss M (1982) Can J Chem Eng 60: 384
13. Gibbs SG, Loy DF, Debelak KA, Tanner RD (1987) in: Biotechnology Processes, Ho and Oldshue (eds), AIChE p 6

14. Margaritis A, Wallace JB (1984) Biotechnology 3: 447
15. Van Brunt J (1987) Biotechnology 5: 1133
16. Knight P (1988) Biotechnology 6: 507
17. Merchuk JC (1988) Encontro Latino-Americano de Biotechnologia, OEA, San Paulo (July 1988)
18. Oldshue JY (1983) Mixing Technology, McGraw-Hill, New York
19. Weetman RJ, Oldshue JY (1988) Power, Flow and Shear Characteristics of Mixing Impellers Proc. 6th European Conference on Mixing, Pravia Italy, p 43
20. Oldshue JY (1983) Genetic Engineering News, p 46
21. Metzner AB, Otto RE (1957) AIChE J 3: 13
22. Calderbank PH, Moo-Young M (1961) Trans Inst Chem Engrs 39: 337
23. Ducla JM, Desplanches H, Chevalier JL (1983) Chem Eng Comm 21: 29
24. Stein WA (1986) Chem Eng Process 20, 137
25. Sinskey AJ, Fleischaker RJ, Tuo NA, Giard GJ, Wang DIC (1981) Ann. NY, Acad Sci 369: 47
26. Hu WS (1983) Ph. D. Thesis, MIT
27. Crougham MS, Hamel JF, Wang DIC (1987) Biotechnol Bioeng 29: 130
28. Wallis GB (1969) One-Dimensional Two-Phase Flow, McGraw-Hill, N.Y.
29. Rushton JH, Costic EW, Everett HJ (1950) Chem Eng Prog 46: 467
30. Nishikawa M, Kato H, Hashimoto K (1977) Ind Eng Chem Process Des Dev 16: 133
31. Nakano N, Yoshida F (1980) Ind Eng Chem Process Des Dev 19: 190
32. Kawase Y, Moo-Young M (1986) Chem Eng Commun 40: 67
33. Popovic M, Robinson CW (1984) Proceedings of the 34th Canadian Chemical Engineering Congress, Quebec City, p 238
34. Ueyama K, Miyauchi T (1976) Kagaku Kogaku Rombusho 2: 595
35. Ueyama K, Miyauchi T (1979) AIChE J 25: 258
36. Yang Z, Rustmeyer U, Buchholz R, Onken U (1986) Chem Eng Comm 49: 51
37. Molerus O, Kurtin M (1986) Chem Engng Sci 41: 2685
38. Joshi JB, Sharma MM (1979) Trans Instn Chem Engrs 57: 244
39. Zehner P (1986) Int Chem Eng 26: 22
40. Camposano A, Chain EB, Gualandi G (1958) 7th International Congress of Microbiology, Stockholm, Sweden (August 1958)
41. Pitt DE, Bull AT (1982) Trans Br Micol Soc 78: 97
42. Metz B (1976) Ph. D. Thesis, Delft University of Technology, The Netherlands
43. Konig B, Seewald Ch, Schugerl K (1981) Eur J Appl Microb Biotechnol 12: 205
44. Ujcova E, Fend Z, Musilcova M, Seichert L (1980) Biotechnol Bioeng 22: 237
45. Bronnenmeier R, Märkl M (1982) Biotechnol Bioeng 24: 553
46. Märkl H, Bronnenmeier R (1985) Biotechnology, Rehm HJ, Reed J (eds) H. Brauer (vol ed) vol II, Chap 18, VCH, Weinheim
47. Tanaka H, Takahashi J, Veda J (1975) J Ferment Technol 53: 18
48. Reuss M (1988) Chem Eng Technol 11: 178
49. Wase JD, Ratwate AM (1985) Appl Microbiol Biotechnol 22: 325
50. Wase JD, Patel YR (1985) J Gen Microbiol 131: 725
50. Wase JD, Patel YR (1985) J Gen Microbiol 131: 725
51. Yerushalmi L, Volesky B (1985) Biotechnol Bioeng 27: 1297
52. Funahashi H, Maehara M, Taguchi I, Yoshida T (1987) J Chem Eng Japan 20: 16
53. McNeil B, Kristiansen B (1987) Biotechnology Letters 9: 101
54. Silva HJ, Cortinas T, Ertola JR (1987) J Chem Tech Biotechnol 40: 41
55. Dewey CF, Bussolari RS, Gimbrone MA, Davies DF (1981) J Biomech Engin 103: 177
56. Folkman J, Moscona A (1987) Nature 273: 345
57. Hollis TM, Ferrone RA (1974) Experimental and Molecular Pathology 20: 1
58. De Forrest JM, Hollis TM (1980) Experimental and Molecular Pathology 32: 217
59. Stathopoulos NA, Hellums JD (1985) Biotechnol Bioeng 27: 1021
60. Frangos JA, McIntre LV, Eskin SG (1988) Biotechnol Bioeng 32: 1053
61. Petersen JF, McIntre LV, Papoutsakis ET (1988) J Biotechnology 7: 229
62. Handa A, Emery AM, Spier RE (1987) 4th Europeqan Congress on Biotechnology, Neijsell OM, Van der Meer RR, Lyuben KChAM (eds) THP-291, Vol 3 p 601
63. Tanaka H (1981) Biotechnol Bioeng 23: 1203
64. Fowler MW (1982) Prog Ind Microb 17: 207

65. Meijer JJ, Van Gulik WM, Ten Hopen JJG, Luyben KChAM (1987) Proc., 4th European Congress on Biotechnology, Meijssel, van der Meer and Luyben (eds) Vol. 2, Elsevier, Amsterdam, p 409
66. Scragg AH, Allan EJ, Leckie F (1988) Enzyme Microb Technol 10: 361
67. Hixson AW, Crowell JH (1931) Ind Chem Eng 23: 923
68. Tanaka H, Semba H, Jitsufuchi T, Harada H (1988) Biotechnology Letters 10: 485
69. Sherman P (1970) Industrial Rheology: With Particular Reference to Food, Pharmaceuticals and Cosmetics, Academic Press, NY
70. Midler M, Finn KR (1966) Biotechnol Bioeng 8: 71
71. Smith CG, Greenfield PF, Randerson DH (1987) Biotechnol Tech 1: 39
72. Schürch U, Kramer H, Einsele A, Widmer F, Eppenberger HM (1988) J Biotechnology 7: 179
73. Chittur KK, McIntre LV, Rich RR (1988) Biotechnology Prog 4: 89
74. Crougham MS, Wang DIC (1989) Biotechnol Bioeng 33: XXX
75. Wutke M, Schügerl K (1988) Modern Approaches to Animal Cell Technology, Spier RE, Griffiths (eds) Butterworths, London, p 297
76. Edwards N, Beeton S, Bull AT, Merchuk JC (1989) Appl Microbiol Biotech (in press)
77. Merchuk JC, Edwards N (1988) UK Patent Application
78. Young TB (1979) Ann NY Acad Sci 326: 165

Byproducts from *Zymomonas mobilis*

M. R. Johns[1], P. F. Greenfield[1], and H. W. Doelle[2]
[1] Department of Chemical Engineering; [2] Department of Microbiology,
The University of Queensland, St. Lucia Australia 4067

Z. mobilis is a microorganism that is not only an extremely efficient producer of alcohol, but also is capable of producing other metabolites in high concentrations under the correct culture conditions. The technology exists to manufacture fructose, sorbitol and gluconic acid at high yields and rates; all three are valuable chemicals. Levan and fructooligosaccharides represent further metabolites of interest, although the optimal conditions required for their formation are still unclear and their usefulness is yet to be fully established. Furthermore, several enzymes have a potential market and could be extracted from waste cell material using established affinity chromatography methods as byproducts of a *Zymomonas* process. Other minor byproducts are not produced in sufficient quantity to justify commercial interest.

Advances in Biochemical Engineering
Biotechnology, Vol. 44
Managing Editor: A. Fiechter
© Springer-Verlag Berlin Heidelberg 1991

1 Introduction

The anaerobic bacterium, *Zymomonas mobilis*, has been intensively studied over the last decade due to its efficient production of ethanol from glucose, and to a lesser extent from fructose and sucrose. Much of this knowledge has been summarised in recent reviews [1–4].

When the substrate for growth is glucose (up to 300 g l^{-1}), ethanol production in batch culture is typically 94–96% of the theoretical maximum [1]. In continuous culture [1], even higher efficiencies (95–98%) have been obtained at initial glucose concentrations of up to 170 g l^{-1}. These ethanol yields are superior to those obtained by yeast when grown on glucose and are due to the lower biomass yield of the bacterium and the absence of byproducts in greater than trace amounts.

The same conversion efficiencies are not observed when the bacterium is cultured on sucrose or fructose. When fructose is used as substrate, growth is slower, biomass yields are only half those of glucose cultures [1] and ethanol yields are typically only 90% of theoretical [5–7]. This is due to byproduct formation, especially dihydroxyacetone, mannitol and glycerol [5].

The results obtained from the growth of *Z. mobilis* on sucrose or cane syrup media suggest that ethanol yields of 85–90% are typical (Table 1), with ethanol yields falling markedly as the sucrose concentration exceeds 150 g l^{-1} or lower grade sucrose material is used. Using molasses and sugar cane juice as substrate, several *Z. mobilis* strains gave ethanol yields of 75% at best [15] on a 7000 litre scale. The reasons for this decline in efficiency was found to be related to two problems [3], sucrose cleavage to fructose and glucose and fructose uptake into the cell [8] resulting in sorbitol and fructo-oligosaccharide formation. Whereas the sucrose cleavage problem can be overcome [16, 17] through adjustments in preculture preparations and optimization of the batch culture conditions, fructose uptake in media exceeding 150 g l^{-1} or high salt concentrations (molasses) remains to be solved [18, 19].

Table 1. Ethanol yields from sucrose using *Z. mobilis*

Z. mobilis strains	Sucrose concentration (g l^{-1})	Culture mode	Ethanol yield[a] (% theoretical)	Reference
UQM 2716	≤ 200	batch	86–93	8
	> 200	batch	60–80	8
	100–150[b]	batch	87.7–93.9	9
ZM4	100	continuous	88–91	1
	150	continuous	75	1
	250	batch	90.2	1
ZM4F JM1[c]	100–150	continuous	80	10
VTT-E-78082	50–200	continuous	50–95	11
	150	batch	75	12
8 strains	120	batch	78–84	13
Z10	50–150	batch	79–90	14

[a] on the basis of initial sucrose; [b] as cane syrup; [c] flocculent strain

The poorer performance of *Z. mobilis* on sucrose compared to glucose is therefore due to increased by-product formation. Problems in sucrose cleavage to glucose and fructose leads to the formation of the fructose polymer levan [11], whereas a sucrose hydrolysis rate in excess of the bacterium's glucose and fructose uptake rates causes an accumulation of both monomers resulting in sorbitol or fructo-oligosaccharide formation (Fig. 1). These carbon diversions lead to poorer ethanol yields and increasing carbon balance discrepancies [3]. The latter can be observed constantly at sucrose concentrations in excess of 150 g l^{-1} as well as at high salt concentration [18, 19].

Large scale trial experiments using corn [20] and milo [21] as substrate together with the discovery that fructokinase is inhibited by glucose [22] revealed that glucose availability and concentration may decide whether *Z. mobilis* channels its

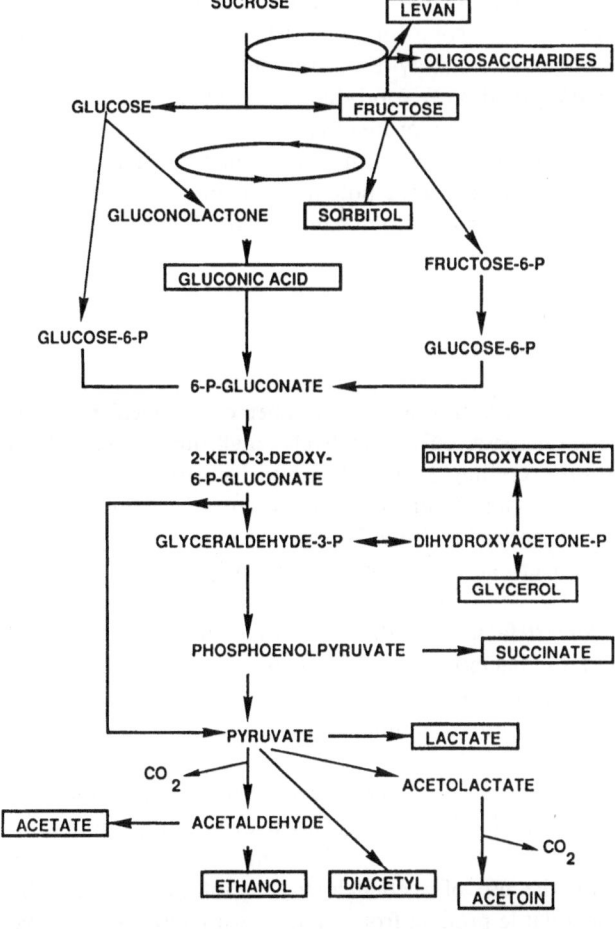

Fig. 1. Metabolic pathways to major byproducts in *Zymomonas mobilis*

carbon totally via the Entner-Doudoroff pathway to ethanol or only partly with
the rest of the carbon forming the above-mentioned byproducts.

Some of these byproducts have considerable commercial value and their
formation by Z. *mobilis* offers a potential new route for their manufacture. The
bacterium is an attractive biocatalyst for industrial scale processes for the following
reasons:
1. high specific activity
2. low biomass yields, resulting in lower quantities for waste disposal
3. ethanol production reduces the risk of contamination
4. relatively simple growth requirements
5. aerotolerant, obviating the need for aeration (particularly useful in biopolymer
   production).

Furthermore, the ability to obtain multiple products from a single process
permits the cost of cell culture to be reduced for each product. Provided that the
conditions required are not too different to those optimal for the products singly,
this is economically advantageous, particularly for products such as ethanol where
profit margins are low, and offers scope for greater flexibility in production.
Recovering valuable byproducts is certainly not novel and is widely practised in
many industries, however the practise of co-production, in which the process is
deliberately optimised for more than one product is less common.

This review summarises knowledge concerning byproduct formation by Z.
mobilis and highlights those processes with potential commercial application.
Culture conditions and techniques reported to maximise byproduct formation are
discussed and competing processes are outlined.

2 Minor Byproducts of *Z. mobilis*

Z. *mobilis* has been reported to produce a large number of chemical products
other than ethanol (Fig. 1). Of these, only a relatively small number have been
produced at a concentration exceeding $1\,g\,l^{-1}$. These include fructose, sorbitol,
gluconate, fructooligosaccharides and levan. Typical concentrations observed for
the minor byproducts are summarised in Table 2. Some are produced to maintain
the redox balance under anaerobic conditions, particularly when sorbitol is
produced [30].

Succinic acid may be an artefact, since the enzymes required in yeast [31]
(Saccharomyces cerevisiae) and bacteria [32, 33] for de novo synthesis of succinic
acid are absent in Z. *mobilis* [34]. Alternatively, it may be formed via succinyl-CoA
as the result of the biotransformation of amino acids present in yeast extract or
flours [16].

Aerobic culture of selected strains of Z. *mobilis* was used to obtain enhanced
concentrations of acetaldehyde [26]. Although acetaldehyde is usually inhibitory
to growth at concentrations as low as $0.5\,g\,l^{-1}$, a maximum concentration
of $4\,g\,l^{-1}$ was achieved by the use of acetaldehyde-tolerant mutants and by
continuous stripping of the volatile product from solution with aeration. Aerobic
conditions also promote the formation of acetate by Z. *mobilis* cultures [25].

Table 2. Minor byproducts of *Z. mobilis*

Byproduct	Initial substrate concentration $(g\,l^{-1})$	Typical byproduct concentration $(g\,l^{-1})$	Reference
Mannitol	fructose (150)	3.7	5
Lactic acid	glucose (10–150)	0.05–1.7	3, 23, 24, 25
	fructose (150)	0.2	5
Acetaldehyde	glucose (40)	4.1	26
	glucose (50–150)	0.1–0.45	3, 23, 24
	fructose (150)	0.3	5
	glu/fru (50 each)	0.5	27
Gluconic acid	glu/fru (50 each)	0.8	25, 27
Succinic acid	glucose (50)	0.05–0.28	23
Acetyl methyl carbinol	glucose (50–150)	<0.05–0.7	3, 24
	fructose (150)	0.5	5
	glu/fru (50 each)	0.3	27
Acetic acid	glucose (50–150)	0.1–0.5	3, 23, 24, 25
	fructose (150)	0.6	5
Glycerol	glucose (50–150)	<0.22	3, 23, 24
	fructose (150)	2.5	5
Dihydroxy-acetone	fructose (150)	5.9	5
	glu/fru (75 each)	1.0	5
1-Propanol	glucose (50–150)	tr–0.1	3, 28, 29
1-Butanol	glucose (50–150)	tr–0.04	3, 28, 29
2-Methyl-1-butanol	glucose (50–150)	tr–0.03	3, 28
Isoamyl alcohol	glucose (50–150)	tr–0.05	28, 29
Pentanols	glucose (50–150)	tr–0.03	28

Increased concentrations of both products appear to be in response to the need for regeneration of reduced cofactors to drive the removal of dissolved oxygen from solution by the NAD(P)H-oxidase reaction during aerobic culture [4, 25]. In respect to the pathway for acetate synthesis, some confusion exists regarding the presence of acetaldehyde dehydrogenase activity in *Z. mobilis* [4, 25, 26].

Fusel oil production by *Z. mobilis* strains is significantly less than that of the yeast, *S. cerevisiae*, with the average total higher alcohols formed in growing cultures of each being 26 mg l^{-1} and 948.8 mg l^{-1}, respectively [34]. This simplifies the distillation process for alcohol recovery from the bacterial culture medium. When grown on glucose, the higher alcohols characteristic of *Z. mobilis* are 1-propanol, 1-butanol and isoamyl alcohol (3-methyl-1-butanol) [28, 34]. De novo synthesis of higher alcohols from glucose in six strains of *Zymomonas* was either not greatly changed or markedly reduced if resting cells were used [28]. The addition of certain amino acids and/or their precursors to the medium of resting cells of *Z. mobilis* 8938 or *Z. mobilis* subsp. *mobilis* B806 increased production of some alcohols by an order of magnitude. It was suggested that bioconversion of the amino acids to alcohols was occurring. Similar results were reported by Bevers and Verachtert [29]. Phenol was identified in the medium of the above strains

when phenylalanine or tyrosine were added to resting cell cultures [28]. Traces of
the ester, ethyl acetate have been identified in cultures of *Zymomonas* grown on
150 g l^{-1} glucose [3]. The enzymes responsible for these conversions, however,
have not been identified to date.

While many of these minor products are of commercial importance, their
production by *Z. mobilis* is too low to be of significance.

3 Fructose

3.1 Use of Fructose and Present Technologies

Fructose is a naturally occurring monosaccharide with a sweetness greater than
that of any other sugar and, in the most common β-fructopyranose form, has a
sweetness 1.7 times greater than sucrose. The primary commercial application of
fructose is as a calorific sweetener. The production of fructose as High Fructose
Corn Syrup (HFCS) in the USA over the last decade is given in Table 3. The
USA has produced 75% of the world output of HFCS during this period [35, 36].

The predominant technology for the large scale manufacture of fructose involves
the conversion of corn starch to fructose using starch hydrolysing enzymes to
produce a 95% glucose syrup followed by the use of immobilised glucose isomerase
to generate a fructose/glucose mixture that can be chromatographically enriched
and blended to give fructose-rich syrups for use in the food industry. This
technology has been reviewed by a number of authors [37–41].

A recent report demonstrated that the use of glucose isomerase in a non-aqueous
environment (85–90% ethanol) favourably affected the reaction equilibrium to
achieve concentrations of 55% fructose from glucose syrup and offered the
possibility of eliminating the chromatographic enrichment step [42].

Table 3. Production of HFCS in the USA during 1980–1989 [35]

Year	Fructose production (1000 short tons, dry weight)		
	HFCS-42	HFCS-55	Total
1980	1530	650	2180
1981	1603	1069	2672
1982	1554	1554	3108
1983	1622	1982	3632
1984	1610	2684	4294
1985	1825	3388	5213
1986	1872	3485	5357
1987	2035	3610	5645
1988	2303	3555	5858
1989 (estd)	2535	3500	6035

HFCS-55 syrup comprises 55% fructose, 41% glucose, 4% higher saccharides (dry basis).
HFCS-42 syrup comprises 42% fructose, 50% glucose, 8% other saccharides

Fructose syrups can also be generated from the hydrolysis of inulins, which are linear β-2,1-linked fructose polymers with a terminal glucose unit [40]. Extensive research is underway to source inulin-rich plants such as Jerusalem artichokes and chicory and to develop hydrolysis processes [39, 40, 43–45].

Fructose can be obtained from cane sugar by the inversion of a sucrose syrup and subsequent chromatographic resolution of the fructose from glucose [46–48]. This has been the basis for the manufacture of high quality, crystalline grades in the past. An earlier process eliminated glucose from invert syrup by the use of glucose oxidase to produce gluconic acid, which was separated from fructose by precipitation [49]. A related concept was developed by Cetus Corp with the enzymic conversion of glucose to glucosone (2-keto-glucose) [50]. The glucosone can be subsequently hydrogenated to fructose. However it is doubtful whether these processes can compete with the conventional HFCS route.

A different approach to obtain fructose from sucrose is the use of the fructosyl transferase enzyme of *Aureobasidium pullulans* via fructose polymer [51] or the concomitant production of ethanol and fructose using fungi [52, 53], yeast [54, 55] or *Zymomonas mobilis* [8]. The latter approach has several attractive features:
1. The separation of fructose from ethanol is straightforward in contrast to its separation from glucose.
2. Both chemicals have large markets and fructose is more valuable than ethanol.
3. *Z. mobilis* is particularly efficient in producing ethanol from glucose, but significantly less so from sucrose.
4. *Z. mobilis* possesses two separate enzymes for monomeric sugar utilization, glucokinase and fructokinase. Consequently, fructokinase-negative mutants, deficient in fructose utilization, can be produced, which only form ethanol from glucose.

3.2 Production of Fructose by *Z. mobilis*

3.2.1 From Sucrose

The disaccharide sucrose is comprised of the two monosaccharides, glucose and fructose, linked by an α-1,2-glycosidic bond. In batch cultures of wild-type *Z. mobilis* strains grown on media containing sucrose at concentrations less than 150 g l^{-1}, complete conversion of sucrose to ethanol and other products occurs and fructose accumulation is transient (Fig. 2). Despite the passage of a decade since the review of Swings and De Ley [34], it remains unclear which enzymes participate in the hydrolysis of sucrose. The enzyme levansucrase (β-2,6-fructan : D glucose-1-fructosyl transferase, EC 2.4.1.10) is considered to be largely responsible and appears to occur in several forms, as a cell-associated [56], intracellular [3] and extracellular enzyme [57, 58]. Although the presence of an invertase-type enzyme has been demonstrated as extremely doubtful [18, 19], other proteins exhibiting sucrase activities, that is sucrose-hydrolyzing without levan producing capability, have been found in culture media of *Z. mobilis* ATCC 31821 (strain ZM4) [58].

Under batch culture conditions, the rate of sucrose hydrolysis depends entirely on the sucrose concentration, pH and temperature of the medium for a given

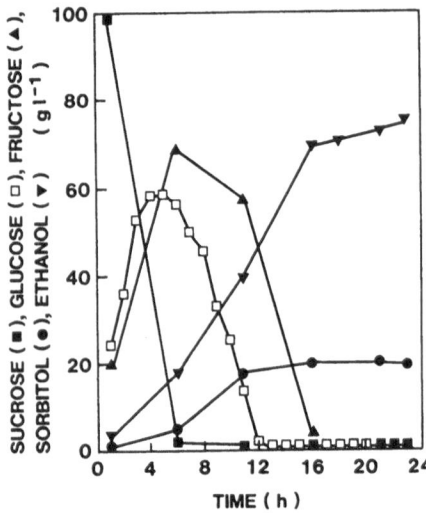

Fig. 2. Ethanol and byproduct formation from sucrose in a batch culture of *Z. mobilis*

strain and is totally independent of the monomeric glucose and fructose uptake rates. When the sucrose hydrolysis rate exceeds the monomeric sugar uptake rates (e.g. above $100\,\mathrm{g\,l^{-1}}$ sucrose concentrations) transient accumulation of the monosaccharides occurs in the medium shortly after inoculation. At this time, glucose and fructose concentrations may exceed $20\,\mathrm{g\,l^{-1}}$, depending on the strain used and the initial sucrose concentration. The disparity between the rates of sucrose hydrolysis and uptake of the monosaccharides is increased with increasing sucrose concentration and increasing dilution rate in continuous culture [11] (Fig. 3).

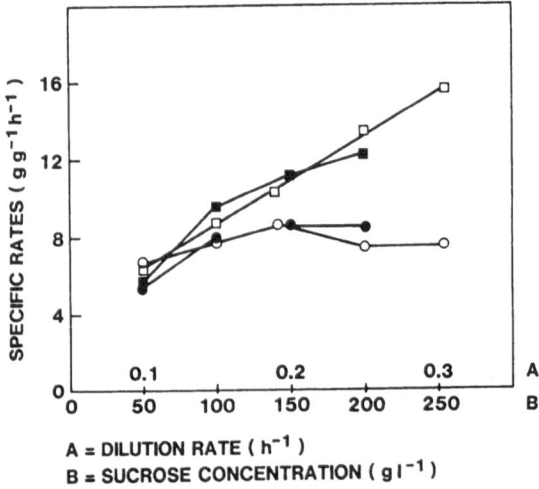

A = DILUTION RATE (h^{-1})
B = SUCROSE CONCENTRATION ($g\,l^{-1}$)

Fig. 3. The disparity between rates of sucrose hydrolysis (□, ■) and monosaccharide uptake (○, ●) continuous culture of *Z. mobilis* [11]: *Open symbols* — dilution rate; *closed symbols* — sucrose concentration

Since no facilitated diffusion or active transport of sucrose has yet been observed in *Z. mobilis*, sucrose is first hydrolyzed to glucose and fructose outside the cell, which in turn agrees with the high sucrase activities found in the culture medium.

Glucose and fructose are carried into the cell by a carrier-mediated diffusion transport system [59] and simultaneously catabolized via the Entner-Doudoroff pathway [14]. The initiation of catabolism occurs by phosphorylation of both sugars at the 6-position using two substrate specific and constitutive enzymes, glucokinase (EC 2.7.1.2) and fructokinase (EC 2.7.1.4) [60].

Whereas glucokinase was found to be under nucleotide control [22], fructokinase appears to be regulated mainly by the internal glucose acccumulation. The latter phenomenon together with the observation in fructokinase-negative mutants [61, 62] that glucokinase is able to phosphorylate fructose at high fructose concentrations indicate that monomeric sugar transport and its subsequent phosphorylation must follow different regulatory control systems. Such a different control of transport and catabolism allows the accumulation of either or both monomeric sugars inside and outside the cell under conditions whereby sucrose hydrolysis exceeds the catabolic activity. This in turn supports the theory that catabolic enzymes or metabolic rates determine the rate of monomeric sugar utilization [63]. Consequently, it is the catabolic activity and not the sucrose hydrolyzing activity which limits the rate of ethanol formation.

The independence of the sucrose hydrolyzing activity and monomeric sugar transport from the rate of monomeric sugar utilization, together with the different control mechanisms acting on the two initial phosphorylating enzymes, can lead to a situation whereby *Z. mobilis* switches to fructose, sorbitol and oligosaccharide formation (Fig. 1).

This phenomenon can be observed when batch cultivations using initial sucrose concentrations up to 150 g l^{-1} are compared with those in excess of 150 g l^{-1}. At 100 g l^{-1} sucrose, the conversion of sucrose to ethanol is complete within $10-12$ h with negligible byproduct formation. At 150 g l^{-1} sucrose, the transient accumulation of glucose and fructose leads to the formation of sorbitol (Fig. 2). In batch cultures of *Z. mobilis* UQM 2716 at initial sucrose concentrations increasing from 288 to 400 g l^{-1}, glucose was always completely metabolized whereas residual fructose concentrations increased representing 12% of the initial fructose (as sucrose) at 288 g l^{-1} sucrose to 60% at 400 g l^{-1} sucrose [64]. The medium also contained variable amounts of unhydrolyzed sucrose and sorbitol at 400 g l^{-1} initial sucrose. At even higher sucrose concentrations, glucose remained despite culturing times of up to 96 h.

Although there are suggestions [10, 65] that the accumulation of fructose and other byproducts may be due to ethanolic inhibition occurring at ethanol concentrations of $65-75 \text{ g l}^{-1}$, it is now believed that the phenomenon of carbon regulation, described above, must be the factor responsible, since *Z. mobilis* is known to produce more than 75 g l^{-1} ethanol without significant byproduct formation [17]. The hypothesis of carbon regulation is further supported by results from continuous cultures [10, 11] using sucrose as substrate, where fructose accumulates at dilution rates above 0.15 h^{-1} and the medium contains considerable quantities of glucose and other byproducts, particularly levan, fructooligosaccharides and sorbitol. These conditions are plainly unfavourable for the production of ethanol and fructose from sucrose in high yields.

The most significant improvement in ethanol and fructose yields from sucrose have come from the use of mutant strains of *Z. mobilis* lacking fructokinase activity. The development of these mutants made use of the discovery that glucose and fructose were phosphorylated by two separate enzymes [60]. The construction of such mutants was successfully achieved by two research groups using different mutagenic agents, ethylmethane sulfonate (EMS) [66] and nitrosoguanidine (NTG) [27], and the parent strains *Z. mobilis* ATCC 39 676 and ATCC 29 191, respectively. All three strains obtained, ATCC 53 431, ATCC 53 432 [66] and DSM 3126 [27], were found to be fructokinase-negative.

Batch cultivation comparisons between the fructokinase-negative strains and their respective parent strains revealed that they did not differ in their glucose utilisation pattern nor in respect to sorbitol formation with sucrose or glucose/fructose mixtures as sole carbon and energy source. In batch cultures of these mutant strains, high concentrations of fructose have been obtained (Table 4) [27, 66]. The fructose yield is reduced, however, by the formation of large quantities of sorbitol. Furthermore at the high initial sucrose concentrations used, some sucrose remained unhydrolysed [66].

When ethanol and fructose are the desired products, sorbitol formation is undesirable for several reasons:

1. It lowers fructose yields (Fig. 1).
2. It is extremely difficult to separate from fructose [67].
3. Stringent regulations governing sorbitol consumption exist in many countries due to problems of gastric intolerance of this substance in humans [68].

The suppression of sorbitol formation by wild-type *Z. mobilis* in batch cultures using sucrose media has not been achieved. Better results have been obtained by using fructokinase-negative mutants and/or fed-batch or continuous culture methods.

Table 4. Fructose yields from sucrose cultures of *Z. mobilis* mutants lacking fructokinase activity

F mutant	Concentration				% Yield		Reference
	sucrose $(g\,l^{-1})$	fructose $(g\,l^{-1})$	sorbitol $(g\,l^{-1})$	ethanol $g\,l^{-1})$	fructose (% theoretical [a])	ethanol	
1959	200	82.1	25.2	43.1	78.0	80.0	27
E 977	203	83.2	17.7	52.1	82.3	101.0	66
E 977	217	109.7	4.1	51.7	103.0	95.2	66
E 977	390	171.7	34.5	69.4	89.8	71.2	66
2864	350[b]	142.0	16.0	76.5	80.6	86.2	62
	350[b]	118.2	7.1	81.8	64.2	90.5	62

[a] based on sucrose consumed. [b] fed-batch technique

The accumulation of glucose in the aqueous phase during growth is responsible for sorbitol formation. To prevent this, sucrose hydrolysis must be synchronised with glucose uptake, since both events are regulated independently of each other [14]. This can be achieved by pH control of the culture. In using sugar cane syrup containing 100 g l^{-1} sucrose, it was found [69] that a pH control between 5.7 and 6.0 favoured fructose recovery. This is in agreement with earlier observations that levansucrase activity and thus sucrose hydrolysis is optimal around pH 6.0 [56] and ethanol production between pH 4.5–5.5 [10, 70].

Bringer-Meyer and co-workers [26] suggested that continuous cultivation of fructokinase-negative mutants may lead to synchronization between sucrose hydrolysis and glucose uptake and therefore reduce sorbitol formation, but no experimental data were presented to support this view.

In recent continuous culture studies [69] using a 100 g l^{-1} sucrose-containing semi-defined medium, and using mutant strain UQM 2864 (ATCC 53 431) continuous feeding was started in the late exponential growth phase (10 h) of a batch culture. However, as soon as the dilution rate exceeded 0.1 h^{-1}, unhydrolyzed sucrose appeared in the medium resulting in a sharp decrease in product formation, although the cell yield stayed relatively constant up to 0.2 h^{-1}.

The system was changed by increasing the sucrose concentration in the feed to 250 g l^{-1} with a constant dilution rate set at 0.05 h^{-1} and pH controlled at 6.0. This technique led to fructose levels of 89 g l^{-1} within 48 h and sorbitol levels as low as 3 g l^{-1}. Unfortunately this fructose/sorbitol ratio of 30 could not be maintained [61] as the mutant strain was not capable of maintaining its high sucrose hydrolysing activity. The mutant never reached true steady state conditions.

Significant improvements in the performance of fructose-negative mutants have been obtained by the use of fed-batch culture techniques [62, 69] (Table 4). In fed-batch cultures of the fructose-negative mutant *Z. mobilis* ATCC 53 431 grown on diluted sugar cane syrup, high fructose and ethanol yields were obtained with minimal formation of sorbitol and complete sucrose utilisation. This technique allowed the synchronization of sucrose hydrolysis and glucose uptake [71] as free glucose levels could be maintained at a relatively constant level not exceeding 7 g l^{-1}. Using sucrose feeds of 250 g l^{-1}, a 99% sucrose hydrolysis and an 80% theoretical fructose accumulation was obtained with a fructose/sorbitol ratio of 33.7 [61, 69]. In increasing the sucrose concentration in the feed it was observed that the synchronization between sucrose hydrolysis and glucose uptake is strongly dependent on the initial free glucose level prior to commencement of the nutrient feeding. Scale up of the process to 100 litre cultures returned similar results (Fig. 4).

The use of low initial sucrose concentrations and the controlled feeding of a concentrated syrup once the glucose concentration has fallen below $5-7 \text{ g l}^{-1}$ ensures that glucose levels are kept low throughout the culture circumventing carbon diversion and activation of the glucose-fructose oxidoreductase responsible for the reduction of fructose to sorbitol [30].

The use of highly concentrated sucrose or cane syrup feeds ($550-600 \text{ g l}^{-1}$) was reported to result in high residuals of sucrose and glucose and sorbitol formation [62]. This could be ameliorated by supplementation of the feed with salts, although product yields were lower than those using lower total sucrose concentrations.

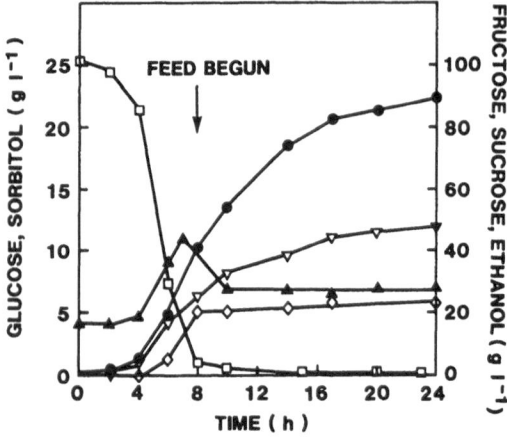

Fig. 4. Fructose production from cane-syrup in a fed-batch culture of fructose-negative *Z. mobilis* UQM 2864 [62]: sucrose (□); glucose (▲); fructose (●); ethanol (▽); sorbitol (◊)

3.2.2 From Glucose/Fructose Mixtures

Relatively little has been reported concerning the production of ethanol and fructose from glucose/fructose mixtures by *Z. mobilis*, although it might be expected that the results should not differ markedly from those obtained with sucrose, since sucrose hydrolysis is rarely rate-limiting. The use of a fructose-negative mutant F-1959 led to poor fructose yields, largely due to high sorbitol levels [27]. The problem was exacerbated at higher sugar concentrations, probably due to the presence of high glucose concentrations, and also led to poor ethanol yields. It is likely that the use of fed-batch methods in which glucose concentrations in the medium are controlled at a low level would improve the results.

3.3 Recovery of Fructose

A number of methods can be used to obtain pure fructose from aqueous solutions. These include aqueous crystallisation [72, 73], alcoholic crystallisation [74, 75], complex formation with calcium [76] and chromatographic techniques [37, 76]. The choice depends largely on the nature of the other chemical species present. Where glucose or other carbohydrates (e.g. sorbitol) are present in even moderate concentrations (above 5–10% on dry sugar basis), crystallisation is ineffective due to the extremely high solubility of fructose (374 g per 100 ml water at 20 °C [77]) in water compared to other sugars. Expensive chromatographic operations, using ion exchange or adsorption resins as in the case of HFCS manufacture, must be used to separate the sugars.

   The use of fed-batch techniques with fructokinase-negative mutants grown on sugar cane syrup produced a medium containing fructose and ethanol with minimal other carbohydrate impurities [62]. Research has been performed to study the feasibility of using the ethanol produced to recover pure, crystalline fructose by alcoholic precipitation and has shown that the presence of sorbitol and/or glucose at concentrations of up to 5% (w/w) on a dry sugar basis can be tolerated [67]. This is well above the levels of these sugars obtained in the bioreactor.

4 Sorbitol and Gluconic Acid

4.1 Use of Sorbitol and Gluconic Acid and Present Technologies

Sorbitol is a hexahydric polyol produced industrially by the reduction of glucose. It is a valuable chemical intermediate with an estimated annual market of 350000 t.p.a. world-wide, most being used in the manufacture of synthetic vitamin C [78]. Other uses of sorbitol have been reviewed recently by Rapaille [79] and include its formulation as an incipient in the tabletting of pharmaceuticals [78], and as an ingredient in foods and industrial chemicals. Sorbitol is the preferred polyol precursor for the synthesis of condensation products such as alkyd and melamine resins and in the synthesis of polyetherpols for use in polyurethane foams due to its high functionality [80]. There is also interest in incorporating dehydration products of sorbitol into polymers due to their improved thermal and mechanical properties. Some trials have demonstrated the efficacy of sorbitol peracetates as bleaching boosters in detergents [80]. Therefore the demand for sorbitol should remain strong.

Gluconic acid and its salts have a smaller world market than sorbitol [81]. The sodium salt is widely used as a chelating agent in cleaning or descaling preparations, especially those formulated for basic environments [82]. Other applications of gluconic acid include its use in the pharmaceutical industry. The lactone derivative, δ-gluconolactone, is used in the food industry as an acidulant.

Current production of sorbitol on a commerical scale is performed by the chemical reduction of glucose over a nickel catalyst at temperatures of 140–150 °C and high pressures (40–50 bar) [81]. Sorbitol can also be formed by the enzymic reduction of fructose, glucose or sorbose and several enzymes capable of catalysing these reactions have been identified in a wide range of microorganisms other than *Z. mobilis* [83].

Gluconic acid is produced commercially by the oxidation of D-glucose using microorganisms [84] or chemical catalytic processes. Both routes have reported yields of 97% or better [82].

4.2 Production of Sorbitol and Gluconic Acid by *Z. mobilis*

Sorbitol and gluconic acid can be the major byproducts of ethanol formation if sucrose or glucose/fructose mixtures serve as substrates in concentrations in excess of those which would limit the metabolic rate of glucose and fructose in *Z. mobilis*. An accumulation of free glucose inside the cell inhibits the activity of fructokinase [22] and induces the glucose oxidative pathway bypassing the glucose phosphorylation step. This pathway, very common in the *Pseudomonas* family of microorganisms (also users of the Entner-Doudoroff pathway, but strictly aerobic), oxidizes glucose via gluconolactone to gluconic acid using the enzymes glucose dehydrogenase and gluconolactonase (EC 3.1.1.17) [85]. Gluconic acid may accumulate, but normally it is phosphorylated by gluconate kinase [86] to

6-phosphogluconate, which is an intermediate of the Entner-Doudoroff pathway. The uniqueness of this diversion is that it bypasses the two regulatory mechanisms acting on glucokinase and glucose 6-phosphate dehydrogenase.

Although the route of carbon flow is similar to that known as the glucose oxidative pathway, the first enzyme participating in glucose oxidation in Z. *mobilis* was found to be a cytoplasmic glucose-fructose oxidoreductase [30]. This newly discovered oxidoreductase not only oxidizes glucose ($K_m = 30$ mM) but also simultaneously reduces fructose ($K_m = 1.4$ M) to sorbitol [88]. The coupled $NAD^+/NADH^+H$ requirement comes most likely from the acetaldehyde to ethanol reduction step, as fewer carbons flow to ethanol under these conditions. The formation of sorbitol and gluconic acid by Z. *mobilis* requires therefore the presence of both glucose and fructose [86] with glucose in excess of 7 g l^{-1}. Sorbitol formation ceases when glucose concentrations fall below this concentration [87].

Gluconic acid may also be formed in small quantity from the oxidation of glucose by a membrane-bound glucose dehydrogenase, which has been detected in cell-free extracts and has a pH optimum for glucose of 6.5 [88].

There have been no reported attempts to optimize culture conditions for sorbitol production by Z. *mobilis*.

4.2.1 Whole Cells

Sorbitol is one of three main byproducts formed by most strains of Z. *mobilis* from the metabolism of sucrose, the other two being levan and fructooligosac-charides. It now seems clear that sorbitol formation is favoured when the rate of sucrose hydrolysis exceeds the glucose uptake rate. The considerable variability that has been observed among different strains with respect to sorbitol formation can be attributed to differences in their rate of sucrose hydrolysis. In one instance, eight strains of Z. *mobilis* were grown on sucrose medium and the amount of sorbitol formed varied from $1-9$ g l^{-1} after 48 h [13]. A direct correlation between the rate of sucrose hydrolysis and sorbitol formation was observed.

A number of operating parameters have been demonstrated to influence sorbitol formation.

(a) *The mode of culture.* Batch culture of Z. *mobilis* leads to increased levels of sorbitol formation compared to other modes of operation. The use of continuous culture resulted in reduced [11] or no [10, 89] sorbitol formation. Steady state sorbitol concentrations increased as the dilution rate $(0.1-0.3$ h$^{-1})$ or the sucrose concentration $(50-200$ g $l^{-1})$ in the feed was raised [11]. This was correlated to increased levels of fructose and glucose in the medium under these conditions. However, the highest sorbitol level observed was 10 g l^{-1}, well below the maximum values observed for batch cultures.

Fed-batch culture with a fructose-negative mutant has been used to obtain high concentrations of fructose from sucrose while minimising sorbitol formation [62]. Sorbitol concentrations were kept below 7 g l^{-1} at total sucrose concentrations of $150-200$ g l^{-1}. Sorbitol formation only occurred late in the growth phase and ceased during the feed stage, despite the presence of high fructose concentrations. This may be due to the repression of the oxidoreductase responsible for sorbitol formation by the high fructose levels [30] and the low glucose concentrations (generally less than 10 g l^{-1}).

(b) *Sucrose concentration.* The concentration of sorbitol was increased in batch cultures of *Z. mobilis* UQM 2716 by increased concentrations of sucrose [64]. A sucrose concentration of 397 g l^{-1} yielded 86 g l^{-1} of sorbitol in a batch culture with the pH controlled at 5.0. This is the highest level reported from cultures of *Z. mobilis* and represents a sorbitol yield of 24.3% of theoretical. However, in most cases reported sorbitol yield based on sucrose consumed is largely independent of initial sucrose concentration and falls in the range of 10–14% of theoretical.

(c) *Culture pH.* Batch cultures controlled at pH 5 led to concentrations of sorbitol approximately double those observed at pH 6 using high sucrose concentrations despite similar sucrose conversion efficiencies [64]. When culture pH was maintained above pH 6, no sorbitol at all was formed [8]. These data conflict with the reported pH optimum 6.2 for the purified glucose-fructose oxidoreductase [30] and appear to agree with the pH 5.3 value of Leigh et al. [87]. Furthermore a pH of 6 is closer to the reported optimum for levansucrase [56]. The reason for the discrepancy is unclear.

(d) *Inoculum size.* Sorbitol formation was increased by increased inoculum size [62], possibly due to the generation of higher transient concentrations of glucose in the medium during sucrose hydrolysis. A recent report observed that a low inoculum size resulted in low sucrose hydrolysis rates in *Z. mobilis* cultures [9].

Batch cultures of *Z. mobilis* grown on glucose/fructose mixtures resulted in sorbitol yields similar to those observed in batch sucrose cultures [5, 83]. The effect of the glucose to fructose ratio on sorbitol formation has been studied in batch culture but the results were inconclusive due to the different growth rates of the bacterium on the mixtures [83].

A recent study [90] using high cell density (12.7 g l^{-1} dry weight) continuous culture of *Z. mobilis* ZM4F JM1 has produced interesting data. Medium containing equimolar glucose:fructose mixtures (100 g l^{-1} total concentration) was fed to the bioreactor. Unlike continuous cultures grown on sucrose, sorbitol formation increased with dilution rate with significant levels of sorbitol being produced at dilution rates greater than 0.6 h^{-1}, and a maximum of 12.2 g l^{-1} sorbitol formed at $D = 1.2 \text{ h}^{-1}$. At these dilution rates, the fructose concentration increased to above 10 g l^{-1}.

The total concentration of sugar in the feed also had a considerable effect on the concentrations of sorbitol and residual sugars in the effluent (Table 5). Fructose accumulated preferentially to glucose and above 120 g l^{-1} total sugars, the fructose uptake rate fell while that of glucose increased steadily (Fig. 5). A most interesting observation was that steady state sorbitol concentration could be correlated in linear fashion to the ratio of fructose to glucose steady state levels rather than to the individual concentrations (Fig. 6). The authors pointed out that this may account for the low sorbitol formation in continuous cultures of *Z. mobilis* on

112   M. R. Johns, P. F. Greenfield, and H. W. Doelle

Table 5. The effect of total sugar concentration on sorbitol formation by *Z. mobilis* ZM4F JM1 in continuous culture [90]

	Total sugar concentration in feed (g l^{-1})			
	100	120	140	160
Residual glucose (g l^{-1})	2.0	1.5	3.7	14.1
Residual fructose (g l^{-1})	0.2	4.7	22.6	49.4
Sorbitol (g l^{-1})	0.7	6.0	8.2	11.8

sucrose even at high dilution rates (1.2 h^{-1}) and residual sugar levels (20 g l^{-1}) as the fructose:glucose ratio is low (approx. 0.36) due to levan formation. It may also account for the observation that high sucrose hydrolysis rates favour sorbitol formation in batch cultures, since it might be expected that under these conditions the fructose:glucose ratio would also be high.

A process for the production of sorbitol from glucose/fructose mixtures using a fructose-negative strain of *Z. mobilis*, DSM 3126, has been patented [91].

4.2.2 Treated Cell Systems

Toluene-treated cells of *Z. mobilis* ZM4 have been used to generate sorbitol and gluconate as products from equimolar glucose/fructose feeds [81, 84]. Near stoichiometric yields (96–97%) of sorbitol and gluconate were obtained within 15–20 h in a batch, free cell system containing initial concentrations of each sugar in the range 100–300 g l^{-1} and a cell concentration of 25–40 g l^{-1} cell dry weight.

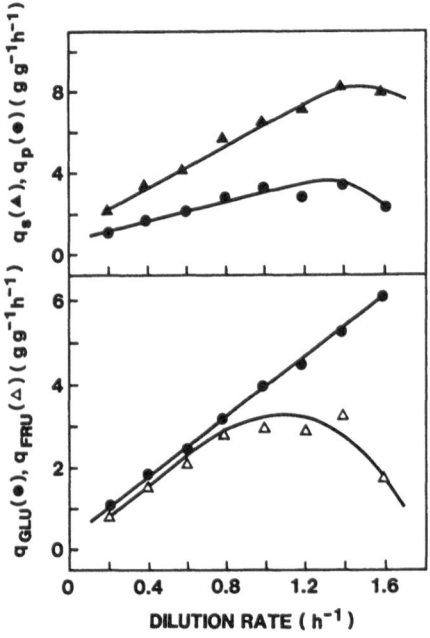

Fig. 5. Effect of dilution rate on the kinetic activity of *Z. mobilis* grown on an equimolar solution of glucose and fructose [90]

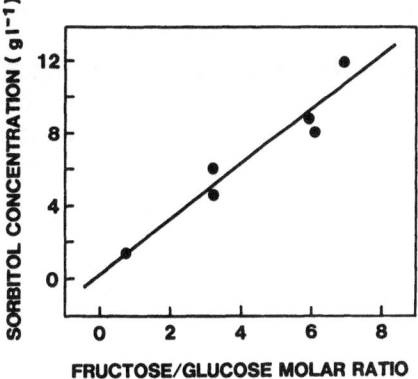

Fig. 6. Effect of fructose to glucose ratio on sorbitol formation [90]

The toluene permeabilised the cells preventing the metabolism of the gluconate formed through the Entner-Doudoroff pathway. The process has been tested in the continuous mode by immobilising the cells in calcium alginate beads [81] and in hollow fibre ultrafiltration modules [92] with similar conversion efficiencies and rates and the technology has been patented [93]. Separation of the two products can be achieved by ion exchange. This process has clear advantages over any culture process, none of which accumulate gluconic acid to any extent.

5 Levan

Levan is a β-2,6-linked fructose polymer with some β-2,1-linked branches formed in *Z. mobilis* cultures as a byproduct of sucrose hydrolysis in which fructose or polyfructose is the electron acceptor. Recent analysis of levan isolated from *Z. mobilis* cultures determined a molecular weight of the order of one million Daltons [94].

Levan formation has not been reported in cultures grown on glucose/fructose mixtures or on the monosaccharides individually [95]. For this reason, it has been suggested that ethanol production from sucrose using *Z. mobilis* may be superior if the sucrose is first inverted to avoid levan formation [90]. The enzyme responsible for levan formation in *Z. mobilis* is levansucrase [95]. This enzyme is induced during the lag and growth phase and also catalyses the rapid hydrolysis (up to $23 \text{ g l}^{-1} \text{ h}^{-1}$) of sucrose [56] into fructose and glucose, although other sucrases may be present in *Z. mobilis*. Glucose and ethanol both inhibit purified levansucrase activity completely at 30 mM (5.4 g l^{-1}) and 1.6 M (73.6 g l^{-1}), respectively [56]. This represents a glucose concentration well below that observed in batch cultures containing high sucrose concentrations. Fructose has no inhibitory effect and may enhance the formation of levan [96]. Although the optimum pH of the pure levansucrase is pH 6.5 [56], optimal production of levan in sucrose cultures is at considerably lower pH [10, 96, 97]. Levan hydrolysis by the bacterium occurs at pH values higher than 5 [96].

Experimental reports available on levan formation show a certain trend which may lead towards the explanation that Z. *mobilis* uses levan formation for the protection of its cellular membrane in a similar way that other microorganisms form capsules or spores for survival. Levan formation has predominantly been observed

1. at low temperatures between 25 and 32 °C or lower [96–98],
2. at low initial sucrose hydrolysis rates, visible by extensive lag phases owing to the requirement for levansucrase induction (insufficient preculture preparation).
3. at high dilution rates (>0.1 h^{-1}) in continuous culture systems [10, 11, 99].
4. at low pH values, e.g. pH 4 [10, 96, 97].

Any one of these conditions is suboptimal for a rapid flux of carbon from sucrose to ethanol and carbon dioxide. High dilution rates exceed the thermodynamic capability of the organism, since Z. *mobilis* prefers growth and energy uncoupling for its ethanol production. Levan can therefore be formed if conditions for growth and metabolism become unfavourable (e.g. high sucrose concentrations), resulting in a severe reduction of sucrose hydrolysis rate.

The rate of sucrose hydrolysis is particularly important in determining the yield of levan. An inverse correlation between the rates of sucrose hydrolysis and levan formation has been observed in shake culture of eight strains of Z. *mobilis* [13] (Fig. 7). High rates of sucrose hydrolysis lead to the transient accumulation of monosaccharides, which promote sorbitol formation from fructose and, in the case of glucose, inhibits purified levansucrase activity at concentrations greater than 5.4 g l^{-1} [56]. Consequently, in batch cultures under these conditions, only low concentrations of levan (2.7 g l^{-1} or less) are typically observed [12].

Continuous cultures, in which the rate of sucrose hydrolysis is generally below that of batch cultures, produce higher levels of levan [10, 11, 92, 95] with

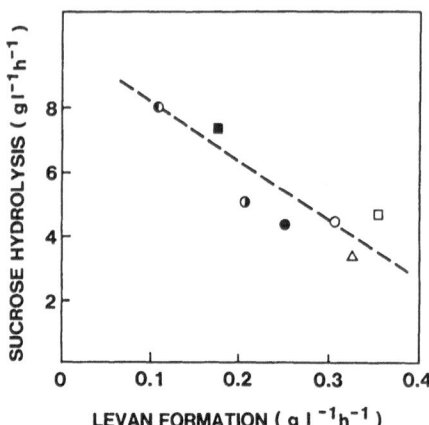

Fig. 7. Relationship of sucrose hydrolysis and levan formation for several Z. *mobilis* strains [13]. The rates are calculated from the period of maximum sucrose hydrolysis rate. The symbols represent different strains

10–17% of the original sucrose (150 g l^{-1}) being converted into levan. The use of a continuous reactor containing *Z. mobilis* immobilised in polyurethane foam and fed 100 g l^{-1} sucrose also resulted in the formation of considerable quantities of levan [89]. A precondition for levan formation in continuous culture is that fructose is not growth-limiting [98].

Recent publications have reported levan concentrations as high as 43 g l^{-1} in batch [97] and continuous culture [96]. These compare well to those for other microbially-produced polysaccharides, such as xanthan gum (approximately 30 g l^{-1}) [100]. *Z. mobilis* would be an attractive microorganism for polysaccharide production, since it does not require oxygen for growth or levan production. Consequently, the problems of heat and mass transfer that limit aerobic biopolymer processes would be circumvented to a large degree.

6 Fructooligosaccharides

Fructooligosaccharides (FOS) are produced as a result of transfructosylation reactions during the growth of *Z. mobilis* on sucrose. Unlike levan, FOS are too small to be precipitated by the amounts of ethanol normally used for levan recovery. Consequently, their presence in cultures of *Z. mobilis* was not suspected in earlier work.

Few quantitative data have been reported concerning FOS synthesis in *Z. mobilis* cultures. In one report [13], the concentration of FOS in shake flask cultures of various strains of *Z. mobilis* grown on 120 g l^{-1} sucrose medium was investigated and found to be constant within the range of 7.6–8.6 g l^{-1}. In contrast, the levan concentration was lower (1.7–2.4 g l^{-1}), except for one culture. Similar results were obtained in a bioreactor with the pH controlled at pH 5 [12].

The FOS produced by *Z. mobilis* are neosugars − short chain FOS with a degree of polymerisation of 3–5, and containing one glucose residue [3]. Since the addition of invertase reverts the polymers to fructose, it is very likely that the oligosaccharides must be of the 1^F- and 6^F-β-fructosyl series as described by Viikari [3].

Most recent observations [17] have shown that the impairment of fructose uptake caused by high salt concentrations (e.g. molasses, NaCl, K$^+$ etc.) leads to large losses in carbon balance calculations. These losses can be recovered upon invertase addition [18, 19]. These results suggest that increasing salt concentrations divert carbon substrate to fructosylation reactions and FOS formation so as to protect the cellular membrane. Since levansucrase activity does not appear to be affected, the sucrose hydrolysis rate remains high, preventing levan formation. The formation of FOS appears to occur during active ethanol formation and must therefore be regarded as an alternative to levan formation.

It is of interest to note that neither levan nor fructo-oligosaccharide formation has, to our knowledge, ever been reported with mixed cultures of *Zymomonas mobilis* and *Saccharomyces cerevisiae*.

Transfructosylation activity leading to oligomer formation is a property of many microorganisms including yeast, fungi and bacteria [3, 101, 102]. FOS are also found in plant tissue. Commonly, these FOS are comprised of fructose linked to sucrose by β-$(2\rightarrow1)$ glycosidic bonds, although β-$(6\rightarrow1)$ oligomers have been reported.

FOS are currently used as a feed additive for swine and poultry in Japan and are being considered for this use in the USA [103]. They are reported to stimulate weight gain and improve feed conversion efficiencies in these animals, probably by stimulating the growth of *Bifidobacteria* in the gut [104]. A sweetener based on FOS (Neo-sugar) has also been produced by the action of fungal fructosyltransferase enzymes on sucrose [103]. The product is 0.4–0.6 times as sweet as sugar and is not digested or adsorbed by humans.

7 Enzymes

The possibility of using *Z. mobilis* for enzyme production was reviewed in a recent article by Scopes [105], who pointed out that this microorganism was the richest source (on an enzyme content per g basis) of many of the enzymes currently used in diagnostic analysis and research. It has been estimated that in active *Z. mobilis* cells, 30–50% of the cellular protein is comprised of glycolytic enzymes [106].

The approximate concentrations of catabolic enzymes in *Z. mobilis* are given in Table 6 [105]. The concentration of these enzymes reached a maximum after 30 h in *Z. mobilis* cultured anaerobically in 200 g l^{-1} glucose, except for pyruvate

Table 6. Approximate levels of the glycolytic enzymes in *Z. mobilis* cells grown at pH 5.0, 30°, on 150 g l^{-1} glucose [105]

Enzyme	EC number	μmol min^{-1} g^{-1}	mg g^{-1}
Glucokinase	2.7.1.2	1000	3
Fructokinase	2.7.1.4	600	2
Phosphoglucose isomerase	5.3.1.9	800	2
Glucose 6-phosphate dehydrogenase	1.1.1.49	1400	2
Gluconate kinase	2.7.1.12	250	0.4
6-Phosphogluconolactonase	3.1.1.31	2500	0.5
6-Phosphogluconate dehydratase	4.2.1.12	1800	5
2-Keto 3-deoxy 6-phosphogluconate aldolase	4.1.2.14	3000	4
Glyceraldehyde 3 phosphate dehydrogenase	1.2.1.12	3000	25
Phosphoglycerate kinase	2.7.2.3	6000	3
Phosphoglycerate mutase	2.7.5.3	12000	5
Enolase	4.2.1.11	1500	7
Pyruvate kinase	2.7.1.40	4500	12
Pyruvate decarboxylase	4.1.1.1	2800	15
Alcohol dehydrogenase-1	1.1.1.1	3000	2.5
Alcohol dehydrogenase-2	1.1.1.1	10000	7
Glucose-fructose oxidoreductase	1.1.99.x	500	1.2

kinase and glyceraldehyde-3-phosphate dehydrogenase which were observed to peak at 18 h and decline thereafter [107]. Optimal conditions for the production and extraction of these enzymes have been discussed by Scopes [105]. The following enzymes might be considered for commercial production.

(a) *Fructokinase.* Fructokinase is a constitutive enzyme catalysing the phosphorylation of fructose to fructose-6-phosphate. The purified enzyme is a dimeric protein with a subunit size of 28 kDa [108], although an earlier study suggested a larger size (85 kDa) [60]. It is highly specific towards fructose, demonstrating no conversion of glucose, sucrose, lactose, galactose or mannose, and possesses an optimum pH of 7.4 [22]. The enzyme exhibited no inhibition by fructose or sucrose at levels of up to 111 mM, and was not product inhibited [22], but glucose competitively inhibited activity with a K_i of 0.14 mM [108]. The *Zymomonas* fructokinase is the only fructokinase reported to have been purified [106] and could find use in the analysis of fructose. However, its strong inhibition by glucose would restrict its application or require the removal of glucose prior to analysis.

(b) *Glucose-6-phosphate dehydrogenase.* This enzyme catalyses the NAD^+-mediated conversion of glucose-6-phosphate to 6-phosphogluconolactone. There is a considerable market for glucose-6-phosphate dehydrogenase, mainly for the use of glucose determination [105]. Currently the enzyme from *Leuconostoc mesenteroides* is used due to its ability to use NAD and NADP as cofactors, the former being considerably cheaper to use in enzyme assays.

The glucose-6-phosphate dehydrogenase from *Z. mobilis* has a tetramic structure with a subunit size of 52 kDa [108]. Like the enzyme from *L. mesenteroides*, it can use either NAD or NADP as cofactor [109]. The purified enzyme has an optimal pH of 8.0 and is not inhibited by glucose concentrations as high as 1 M [108].

Some advantages for the use of the *Zymomonas* enzyme include its requirement for NAD as cofactor, the easier extraction of the enzyme from *Zymomonas* compared to *Leuconostoc*, and the less fastidious nature of the former microorganism [105]. A procedure for the recovery of the enzyme from *Z. mobilis* using Scarlet MX-G dye affinity chromatography with NAD^+ elution has been developed at bench scale [108].

(c) *Pyruvate kinase.* Pyruvate kinase catalyses the generation of pyruvate and ATP from phosphoenolpyruvate and ADP respectively. The *Zymomonas* enzyme is relatively unusual in that it is not controlled by allosteric effectors and is a dimeric rather than a tetramic enzyme, although the subunit size (52 kDa) is similar to other pyruvate kinase enzymes [110]. It has similar specific activity to the commercially available preparations from rabbit muscle and has no requirement for metal cations [110], Scopes [105] suggests that it could be produced at a price competitive with the rabbit muscle preparation.

(d) *Pyruvate decarboxylase (PDC).* Pyruvate decarboxylase from *Z. mobilis* is a hydrophobic, thermostable protein comprising up to 5% of the total protein of the bacterium [111]. The enzyme catalyses the reaction of pyruvate to acetaldehyde with the release of carbon dioxide. The gene for this enzyme has been cloned recently into *Escherichia coli* and expressed at high levels [111–113]. Some variations in the primary structure of the enzyme have been observed between different strains of *Z. mobilis* [114].

Currently there appears to be little commercial application for the purified PDC and only the yeast enzyme is available [105]. It, however, is difficult to purify due to protease attack [112]. In contrast *Zymomonas* PDC is easy to purify and is structurally similar to the yeast enzyme but possesses superior kinetic properties [112]. A number of research groups have cloned the PDC gene from *Z. mobilis* into *E. coli* [115, 116] or *Klebsiella planticola* [117] and obtained significant ethanol production from glucose and pentose sugars, particularly xylose, using these recombinant bacteria. It was shown, in one instance, that the resulting recombinant strain of *E. coli* had improved growth characteristics due to reduced acid formation [116].

(e) *Glucose-fructose oxidoreductase.* Glucose-fructose oxidoreductase is responsible for the formation of sorbitol and gluconolactone from fructose and glucose respectively, using tightly bound NADP as cofactor. The purified enzyme has a pH optimum of 6.2 and the highest yield is obtained by culture of *Z. mobilis* on 200 g l^{-1} glucose medium [30]. The enzyme has a tetrameric structure with a sub-unit size of 40 kDa and appears to be unique to *Z. mobilis*.

The enzyme permits the manufacture of sorbitol and gluconolactone from high concentrations of sugars in high yield and exhibits little product inhibition [30]. A recovery process based on dye affinity chromatography has been developed [104] and several methods using toluene-treated immobilised cells have been reported [81, 92] including one patent [93].

A number of other less well characterised *Zymomonas* enzymes might also be condidates for commercial application including gluconolactonase (EC 3.1.1.17) [30], levansucrase and alkaline phosphatase (AP) (EC 3.1.3.1). The last is not repressed by high phosphate concentrations [118], in contrast to *E. coli* AP. The gene for AP has been recently cloned and expressed in *E. coli* [119].

The greatest challenge facing the use of *Z. mobilis* for enzyme manufacture is the very low cell yield obtained. This is typically in the range 0.01–0.15 g g^{-1} substrate compared to cell yields for many aerobic microorganisms of 0.5 g g^{-1}. For this reason enzyme extraction may only be economic for enzymes unique to *Zymomonas* or where the cell mass is surplus to the large scale production of ethanol. In the latter instance, however, the use of the cells for enzymes would compete with the normal practice of using the cell mass in conjunction with other insoluble medium materials as distillers dried grains for stockfeed.

8 Conclusions

The lower efficiencies of ethanol production from sucrose and glucose/fructose mixtures by *Zymomonas mobilis* compared to that from glucose are due to byproduct formation. The reasons for the byproduct formation can be traced to suboptimal cultivation techniques resulting in a non-synchronization between the sucrose hydrolysis rate and the metabolic rate of the organism. Both rates are completely independent of each other and separated only by a third, rather unknown but increasingly important, control system regulating monomeric sugar and salt uptake via a carrier-mediated diffusion transport system. Since all three systems can be influenced by environmental and/or cultivation technique conditions, sucrose metabolism can be steered towards a 95% ethanol yield or a reduced ethanol yield concomitant with a number of different byproducts, e.g. sorbitol, gluconate, levan or fructo-oligosaccharides. Whereas the conditions for ethanol formation and ethanol plus fructose formation are the furthest advanced, conditions for the economic production of all other byproducts require further intensive research.

9 References

1. Rogers PL, Lee KJ, Skotnicki ML, Tribe DE (1982) Adv Biochem Eng 23: 37
2. Buchholz SE, Dooley MM, Eveleigh DE (1987) Tibtech 5: 199
3. Viikari L (1988) CRC Crit Rev in Biotechnol 7: 237
4. Bringer-Meyer S, Sahm H (1988) FEMS Microbiol Rev 54: 131
5. Viikari L, Korhola M (1986) Appl Microbiol Biotechnol 24: 471
6. Toran-Diaz I, Jain VK, Baratti JC (1984) Biotechnol Letts 6: 389
7. Toran-Diaz I, Delezon C, Baratti JC (1983) Biotechnol Letts 5: 409
8. Doelle HW, Greenfield PF (1985) Appl Microbiol Biotechnol 22: 405
9. Doelle MB, Doelle, HW (1989) J Biotechnol 11: 25
10. Favela Torres E, Baratti J (1987) Appl Microbiol Biotechnol 27: 121
11. Viikari L, Linko M (1986) Biotech Letts 8: 139
12. Viikari L (1984) Appl Microbiol Biotechnol 19: 252
13. Viikari L, Gisler R (1986) Appl Microbiol Biotechnol 23: 240
14. Lyness E, Doelle HW (1981) Biotech Bioeng 23: 1449
15. Schenberg AC, Pinto Da Costa SO (1987) CRC Crit Rev Biotechnol 6: 323
16. Doelle MB, Doelle HW (1989) Aust J Biotechnol 3: 218
17. Doelle MB, Doelle HW (1990) Appl Microbiol Biotechnol 33: 31
18. Doelle MB, Greenfield PF, Doelle HW (1990) Process Biochem Intl. 25: 151
19. Doelle MB, Greenfield PF, Doelle HW (1990) Appl Microbiol Biotechnol 34: 160
20. Doelle MB, Millichip RJ, Doelle HW (1989) Process Biochem 24: 137
21. Millichip RJ, Doelle HW (1989) Process Biochem 24: 141
22. Doelle HW (1982) Eur J Appl Microbiol Biotechnol 14: 241
23. Wecker MSA, Zall RR (1987) Appl Environ Microbiol 53: 2815
24. Schmidt W, Schürgerl K (1987) The Chem Eng J 36: B39
25. Jobses IML, Egberts GTC, van Baalen A, Roels JA (1985) Biotech Bioeng 27: 984
26. Bringer-Meyer S, Scollar M, Sahm H (1985) Appl Microbiol Biotechnol 23: 134
27. Pankova LM, Shvinka JE, Beker MJ (1988) Appl Microbiol Biotechnol 28: 583
28. Rao SC, Jones LP (1987) Acta Biotechnol 7: 209
29. Bevers J, Verachtert H (1976) J Inst Brew London 82: 35
30. Zachariou M, Scopes RK (1986) J Bacteriol 167: 863
31. Oura E (1977) Proc Biochem 12: 19
32. Stanier RY, Ingraham JL, Wheelis ML, Painter PR (1987) General microbiology, 5th edn, MacMillan, New York USA
33. Radler F (1986) Experientia 42: 884
34. Swings J, De Ley J (1977) Bacteriol Rev 41: 1
35. Lord RC, Barry RD, Fry J (1989) Sugar and sweetener situation and outlook report USDA-ERS

36. Hodgkin JA (1987) Sugar y Azucar 82(8): 15
37. Hamilton BK, Colton CK, Cooney CL (1974) In: Olson AC, Cooney CL (eds) Immobilised Enzymes in Food and Microbial Processes, Plenum, NY, p 85
38. Blanchard PH, Geiger EO (1984) Sugar Technol. Rev. 11: 1
39. Bucke C (1981) In: Birch GG, Blakeborough N, Parker KJ (eds) Enzymes and food processing, Applied Sci Pub, London p 51
40. Vandamme EJ, Derycke DG (1983) Adv Appl Microbiol 29: 139
41. Jensen VJ, Rugh S (1987) Methods in Enzymol 136: 356
42. Visuri K, Klibanov AM (1987) Biotech Bioeng 30: 917
43. Kosaric N, Wieczorek A, Cosentino GP, Duvnjak Z (1985) Adv Biochem Eng 32: 1
44. Fleming SE, Grootwassink JWD (1979) CRC Crit Rev Food Sci Nutr 12: 1
45. Fuchs A (1987) Starch 39: 335
46. Boehringer CF & Sons (1967) UK Pat 1,085,696
47. Boehringer CF & Sons (1968) UK Pat 1,117,903
48. Keller BW, Reents AC, Laraway JW (1981) Starch 33: 55
49. Holstein AG, Holsing GC (1962) US Pat 3,050,444
50. Niedleman SL, Amon WF, Geigert J (1981) US Pat 4,246,347
51. Heady RE (1981) US Pat 4,276,379
52. Ueng PP, McCracken LD, Gong CS, Tsao GT (1982) Biotechnol Lett 4: 353
53. Bell JM, Erfle JD, Spencer JFT, Reusser F (1958) Can J Animal Sci 38: 122
54. Duvnjak Z, Koren DW (1987) Biotechnol Lett 9: 783
55. Heady RE (1982) US Pat 4,335,207
56. Lyness E, Doelle HW (1983) Biotechnol Lett 5: 345
57. Mortatte MPL, Sato HH, Park YH (1983) Biotechnol Lett 5: 229
58. Preziosi L, Michel GPF, and Baratti J (1990) Arch. Microbiol 153: 181
59. Di Marco AA, Romano AH (1985) Appl Environ Microbiol 49: 151
60. Doelle HW (1982) Eur J Appl Microbiol Biotechnol 15: 20
61. Edye LA, Kositanont C, Doelle HW (1990) Acta Biotechnol 1: 49
62. Edye LA, Johns MR, Ewings KN (1989) Appl Microbiol Biotechnol 31: 129
63. Cromie S, Doelle HW (1982) Eur J Appl Microbiol Biotechnol 14: 69
64. Doelle HW, Greenfield PF (1985) Appl Microbiol Biotechnol 22: 411
65. Rogers PL, Lee KJ, Tribe DE (1980) Process Biochem 15: 7
66. Johns MR, Greenfield PF (1988) Aust Inst Food Sci & Technol Convention, Manly, p 36
67. Forster H, Mehnert H (1979) Akt Ernahrung 5: 245
68. Suntinanalert P, Pemberton JP, Doelle HW (1986) Biotechnol Lett 8: 351
69. Kositanont C, Edye LA, Doelle HW (1990) Microbios 61: 169
70. Lawford H, Holloway P, Ruggiero A (1988) Biotechnol Lett 10: 809
71. Kositanont C, Edye L, Doelle HW (1989) Proc 8th Aust Biotechnol Conf Sydney, p 247
72. Forsberg KH, Hamalainen L, Melaga AJ, Virtanen JJ (1975) US Pat 3,883,365
73. Shiau LD, Berglund KA (1987) AIChEJ 33: 1028
74. Dwivedi BK, Raniwala SK (1980) US Pat 4,199,374
75. Day GA (1985) Eur Pat 156,571
76. Lauer VK (1980) Starch 32: 11
77. Bates FJ & Assoc (1942) Natl Bur Std (US) Circ C-440
78. Burgess S (1987) Manufact Chemist 58(6): 55
79. Rapaille A (1988) Starch 40: 356
80. Koch H, Roper H (1988) Starch 40: 121
81. Rogers PL, Chun UH (1987) Aust J Biotechnol 1: 51
82. Roper H, Koch H (1988) Starch 40: 453
83. Viikari L (1984) Appl Microbiol Biotechnol 20: 118
84. Chun UH, Rogers PL (1988) Appl Microbiol Biotechnol 29: 19
85. Hardmann MJ, Scopes RK (1988) Eur J Biochem 173: 203
86. Barrow KD, Collin JG, Leigh DA, Rogers PL, Warr RG (1984) Appl Microbiol Biotechnol 20: 225
87. Leigh D, Scopes RK, Rogers PL (1984) Appl Microbiol Biotechnol 20: 413

88. Strohdeicher M, Schmitz B, Bringer-Meyer S, Sahm H (1988) Appl Microbiol Biotechnol 27: 378
89. Amin G, Doelle HW, Greenfield PF (1987) Biotechnol Letts 9: 225
90. Favela Torres E, Baratti J (1987) Biomass 13: 75
91. Bringer-Meyer S, Sahm H (1987) Australian Pat 61 085/86
92. Paterson SL, Fane AG, Fell CJD, Chun UH, Rogers PL (1988) Biocatalysis 1: 217
93. Unisearch (1988) US pat 4,755,467, 6pp
94. Kennedy JF, Stevenson DL, White CA, Viikari L (1989) Carbohydr Polym 10: 103
95. Dawes EA, Ribbons DW, Rees DA (1966) Biochem J 98: 804
96. Reiss M, Hartmeier W (1989) Chem Mikrobiol Technol Lebensm 12: 1
97. Mezbarde I, Pankova LM, Laivenieks M, Svinka J, Bekere M (1989) Latv PSR Zinat Akad Vestis (4) 130
98. Lyness E, Doelle HW (1980) Biotechnol Lett 2: 249
99. Lee KJ, Skotnicki ML, Tribe DE, Rogers PL (1981) Biotechnol Lett 3: 207
100. Gonzales R, Johns MR, Greenfield PF, Pace GW (1989) Process Biochem 24: 200
101. Hidaka H, Eida T (1984) BioIndustry 1: 5
102. Muramatsu M, Kainuma S, Miwa T, Nakakuki T (1988) Agr Biol Chem 52: 1303
103. Fishbein L, Kaplan M, Gough M (1988) Vet Hum Toxicol 30: 104
104. McKellar RC, Midler HW (1989) Appl Microbiol Biotechnol 31: 537
105. Scopes RK (1987) Aust J Biotechnol 1: 58
106. Algar EM, Scopes RK (1985) J Biotechnol 2: 275
107. Osman YA, Conway T, Bonetti SJ, Ingram LO (1987) J Bacteriol 169: 3726
108. Scopes RK, Testolin V, Stoter A, Griffiths-Smith K, Algar EM (1985) Biochem J 228: 627
109. Dawes EA, Ribbons DW, Large PJ (1966) Biochem J 98: 795
110. Pawluk A, Scopes RK, Griffiths-Smith K (1986) Biochem J 238: 275
111. Conway T, Osman YA, Konnan JI, Hoffmann EM, Ingram LO (1987) J Bacteriol 169: 949
112. Neale AD, Scopes RK, Wettenhall REH, Hoogenraad NJ (1987) J Bacteriol 169: 1024
113. Brau B, Sahm H (1986) Arch Microbiol 144: 296
114. Reynen M, Sahm H (1988) J Bacteriol 170: 3310
115. Neale AD, Scopes RK, Kelly JM (1988) Appl Microbiol Biotechnol 29: 162
116. Ingram LO, Conway T (1988) Appl Environ Microbiol 54: 397
117. Feldmann S, Sprenger GA, Sahm H (1989) Appl Microbiol Biotechnol 31: 152
118. Michel GPF, Baratti JC (1989) J Gen Microbiol 135: 453
119. Conway T, Eddy CK, MacKenzie KF, Mejia JP, Pond JL, Utt EA (1988) ASM Ann Meet Abstr

Bioconversions of Ergot Alkaloids

Vladimír Křen

Institute of Microbiology, Czechoslovak Academy of Sciences,
Vídeňská 1083, 14220 Prague, Czechoslovakia

Ergot alkaloids are very important pharmaceutical substances with wide medical use. Their bioproduction is usually followed by chemical modification. Bioconversions of ergot alkaloids on an industrial scale have become more important in recent years. The aim of this study is to collect the available data on ergot alkaloid bioconversions and to show their potential applications. The paper deals with conversions of clavines, e.g. biooxidations, glycosylations of the alkaloids, and bioconversions of lysergic acid derivatives. A part of it is aimed at bioconversions of ergot alkaloids in mammalian organisms and obtaining their metabolites by biotransformation. The use of immobilized systems and cofermentation is also discussed.

Advances in Biochemical Engineering
Biotechnology, Vol. 44
Managing Editor: A. Fiechter
© Springer-Verlag Berlin Heidelberg 1991

1 Introduction

Ergot alkaloids are produced by ascomycetes from the genus *Claviceps*, by some other filamentous fungi and can also be found infrequently in the plant kingdom.

Currently, ergot alkaloids and their derivatives cover a large field of therapeutic uses as drugs of high potency in the treatment of uterine atonia, migraine, orthostatic circulatory disturbances, senile cerebral insufficiency, hypertension, hypergalactinemia, acromegaly and parkinsonism [1].

Ergot alkaloids, like most of the secondary metabolites of microorganisms and plants are produced as a family of related compounds based on the tetracyclic ergoline ring system.

In most of the naturally occurring ergot alkaloids the nitrogen atom at position 6 is methylated and position 8 bears an additional C-atom. In most cases, the ring D carries a double bond in the position 8–9 or 9–10 (8-ergolenes and 9-ergolenes respectively). The asymmetric carbon in the 8-position of the 9-ergolenes gives rise to the two stereoisomeric ergolenes and isoergolenes (Fig. 1). Based on the substituent at C-8, ergot alkaloids are classified as clavines and lysergic acid derivatives. For more details and a complete list of all clavines see reviews by Gröger [2] and Floss [3].

Ring D is seldom open at 6–7 (so-called secoclavines or chanoclavines) or cycled by oxygen.

Ergolene acids are three ergolene-derivatives with a carboxyl group at position 8, i.e. (+)-lysergic acid (**5**), (+)-isolysergic acid (**6**) (9-ergolenes) and paspalic acid (4) (8-ergolenes). The „classical" ergot alkaloids are all amide derivatives of lysergic or isolysergic acid. The so called simple lysergic acid amides are ergine (**7**), lysergic acid α-hydroxyethylamide (**10**), ergometrine (**9**) and their epimers, e.g. erginine (**8**).

The ergot peptide alkaloids consist of a lysergic acid molecule that is linked by an acid amide-type of bond with a cyclole-structured tripeptide. A well known example of this group is ergotamine. For detailed list of these structures see the review by Kobel and Sanglier [4].

Of the naturally occurring ergot alkaloids only two are used in therapy: ergotamine and ergometrine. The rest of the therapeutically important ergot

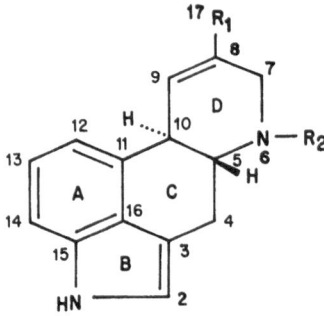

Fig. 1. 8,9-Ergolenes
(*1*) agroclavine $R_1 = CH_3$, $R_2 = CH_3$
(*2*) elymoclavine $R_1 = CH_2OH$, $R_2 = CH_3$
(*3*) noragroclavine $R_1 = CH_3$, $R_2 = H$
(*4*) paspalic acid $R_1 = COOH$, $R_2 = CH_3$

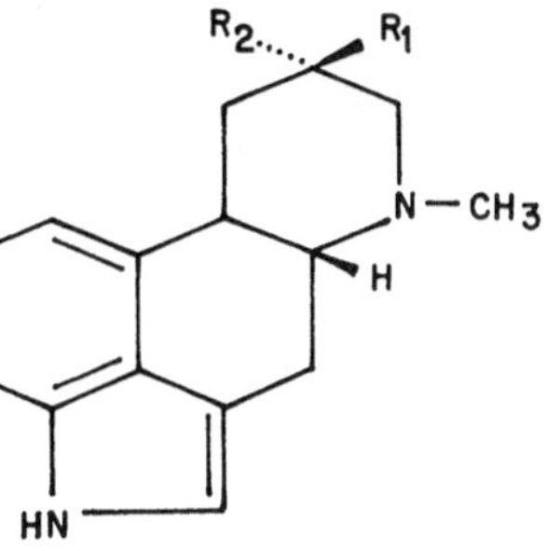

Fig. 2. 9,10-Ergolenes, lysergic acid derivatives
(5) lysergic acid R_1 = COOH, R_2 = H
(6) isolysergic acid R_1 = H, R_2 = COOH
(7) ergine R_1 = $CONH_2$, R_2 = H
(8) erginine R_1 = H, R_2 = $CONH_2$
(9) ergometrine R_1 = $CONHCH(CH_3)CH_2OH$,
    R_2 = H
(10) lysergic acid α-hydroxyethylamide
    R_1 = $CONHCH(CH_3)OH$, R_2 = H
(11) lysergol R_1 = CH_2OH, R_2 = H
(12) iso-lysergol R_1 = H, R_2 = CH_2OH
(13) lysergene $R_{1,2}$ = CH_2
(14) lysergine R_1 = CH_3, R_2 = H
(15) setoclavine R_1 = CH_3, R_2 = OH
(16) penniclavine R_1 = CH_2OH, R_2 = OH
(17) isosetoclavine R_1 = OH, R_2 = CH_3
(18) isopenniclavine R_1 = OH, R_2 = CH_2OH
(29) 8-hydroxyergine R_1 = $CONH_2$, R_2 = OH

compounds have undergone some chemical changes, i.e. elimination of the 9,10-double bond by hydrogenation, halogenation, alkylation etc.

Although the total synthesis of various ergot alkaloids has been demonstrated, it is ruled out for reasons of economy. On the other hand, modification of natural compounds possessing the ergot skeleton has become important. Here, bioconversions can enter with many of their advantages. First, the reaction occurs at chemically "nonactivated" positions. Biological systems exhibit regional selectivity with polyfunctional molecules. This is a great advantage over many chemical reagents that cannot distinguish between multiple similar functional groups. A high degree of stereoselectivity (both substrates and products) is often observed. Workers in this field are rapidly moving toward a more refined approach to choosing organisms because of their predictable type-reactions.

Much of the work on alkaloid transformation was motivated by the need to produce more effective drugs from naturally available compounds. The main effort has been devoted to specific oxidations of ergot alkaloids to produce the desired amount of substrates for semisynthetic preparations. It was demonstrated that many of these reactions are specific for the genus *Claviceps*. These organisms have been widely used for various biotransformations.

Although the alkaloid-producing *Claviceps* strains have been used for many biotransformations, some alkaloids fed to the organisms can be regarded as "xenobiotics". It holds true for the case when the added alkaloids are not naturally produced by the appropriate *Claviceps* strain used (e.g. feeding *C. purpurea* with lysergoles). A similar situation occurs when the converted alkaloid is normally produced as a minor component, and new compounds can be obtained by the "pressure" of its high added concentration (e.g. feeding *C. fusiformis* with a high concentration of chanoclavine). Another effect can arise when high concentrations of the added alkaloid unbalance the regulation of ergot alkaloids synthesis (conversion) towards the desired product as documented by the so called "aggressive bioconversion" [5] (see later). At the moment, only ergot alkaloids bioconversions performed by the strains of the *Claviceps* genus have certain industrial significance and therefore are included in this review.

Some of the studies on ergot alkaloid bioconversions were stimulated by problems of the metabolism of ergot drugs in mammals.

2 Clavine Bioconversions

2.1 Agroclavine

Agroclavine (**1**) can be produced in high yields by selected *C. fusiformis* and *C. purpurea* strains. However, till recently it was not considered as a useful drug. Therefore, the main effort in agroclavine bioconversions was targeted at its oxidation to elymoclavine (**2**) which is an important substrate for semisynthetic ergot alkaloid based drugs. Chemical oxidation to elymoclavine is impracticable.

The enzyme systems from non-*Claviceps* organisms transform agroclavine mostly to 8-hydroxyderivatives. These conversions are catalyzed mainly by peroxidases. The 8-oxidation of 8,9-ergolenes is accompanied by the shift of the double bond to the 9,10-position. Intermediates of this reaction are in some cases 10-hydroxy- or 8,9-epoxy-derivatives.

More than 100 species of filamentous fungi and other microorganisms oxidize agroclavine to setoclavine (**15**) and isosetoclavine (**17**) [6, 7] (Table 1). *Psilocybe semperviva* converts agroclavine with certain degree of stereoselectivity to seco-clavine [8]. The same reaction is performed also by prokaryotic microorganisms as *Streptomycetes* and *Nocardias* [6, 7] and *Pseudomonas aeruginosa* [9].

Table 1. Agroclavine bioconversions

Reaction	Product	Conversion system (culture)	Ref.
8-Hydroxylation	Setoclavine	*Absidia spinosa* ATCC-6648, *Aspergillus carbonarius* PCC-104, *Bispora effusa* CBS, *Cladosporium fulvum* ATCC-10391, *Fusarium solani* PCC-143, *Epicoccum* sp., *Gibberella zeae* Ull, *Helminthosporium carbonum* Ull, *Mucor angulisporus* CBS, *Streptomyces annulatus* PCC-A-111, *S. griseus* PCC, *Nocardia rubra* PCC-252	[6]
		Psilocybe semperviva	[8]
		Penicillium viridicatum	[25]

Table 1. (continued)

Reaction	Product	Conversion system (culture)	Ref.
8-Hydroxylation	Setoclavine + isosetoclavine	*Corticium sasakii*	[17]
		Pseudomonas aeruginosa	[9]
		Horse radish peroxidase	[14, 22]
		Tomato homogenate	[10]
		Potato sprouts homogenate	[10]
		Morning glory seedlings homogenate	[10]
		C. purpurea (more strains)	[18]
2-Hydroxylation	2-Hydroxyagroclavine	*Corticium sasakii*	[7]
10-Hydroxylation	10-Hydroxyagroclavine	Horse radish peroxidase	[12, 13]
8,9-Epoxidation	10-Hydroxy-8,9-epoxyagroclavine	Horse radish peroxidase	[12–14]
17-Hydroxylation	Elymoclavine	*Penicillium roqueforti*	[16]
		Rat liver homogenate	[15]
		Microsomal fraction from *C. purpurea* PEPTY 695/S	[30]
		C. purpurea	[18, 20]
		C. fusiformis	[21, 23–25]
		C. paspali	[5, 25]
		Claviceps sp. strains KK-2 Se-134, 47A, SD-58	[21]
N-6 Demethylation	Noragroclavine	*Streptomyces roseochromogenes, S. punipalus, S. purpurascens*	[7]
		Rat liver homogenate	[15]
		Horse radish peroxidase (+ O$_2$)	[13]
N-6 Demethylation + 8-Hydroxylation	Norsetoagroclavine	Horse radish peroxidase	[12]
8,9-Hydrogenation (not proved)	Pyroclavine + festuclavine	*C. purpurea*	[18]
Agroclavine derivatives 1-Alkyl- 1-Benzyl- 1-Hydroxymethyl- 2-Halo- 2,3-Dihydro- 6-Ethyl-6-nor- 17-Hydroxylation	Corresponding elymoclavines	*C. fusiformis* SD-58	[21]

Peroxidase rich homogenates from tomato fruits, potato sprouts, horse radish and morning glory (Convolvulaceae) seedlings catalyze the oxidation of agroclavine in the presence of H_2O_2 to setoclavine and isosetoclavine [10–12]. Chan-Li and Ramstad [13] found 10-hydroxyagroclavine and 10-hydroxy-8,9-epoxyagroclavine as unstable intermediates of this reaction.

Agroclavine can be also demethylated at the N-6 position by peroxidase under aerobic conditions [12–14]. Mammalian tissue homogenates (rat liver, guinea pig adrenal) produce small amount of both setoclavines from agroclavine [15].

Corticium sasakii [16, 17] produces setoclavines and also 2-hydroxyagroclavine from agroclavine.

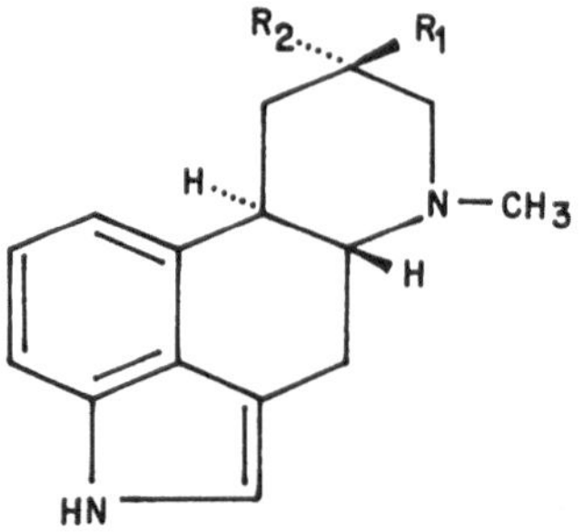

Fig. 3. Ergolines
(*19*) festuclavine $R_1 = CH_3$, $R_2 = H$
(*20*) pyroclavine $R_1 = H$, $R_2 = CH_3$

Some strains of *C. purpurea* are able, beside the main conversion product elymoclavine, to convert agroclavine to setoclavines and to festuclavine (**19**) and pyroclavine (**20**) [18].

The most desired and the most economically important agroclavine conversion is its oxidation to elymoclavine. Although minute amount of elymoclavine was found as agroclavine conversion product in the rat liver homogenate system [15] and by *Penicillium roqueforti* [16], no organism, except for the *Claviceps* strains is able to perform this reaction at a reasonable rate.

Hsu and Anderson [19] found agroclavine 17-hydroxylase activity in *C. purpurea* PRL 1980 cytosol, Kim et al. [20] localized this activity to the microsomal fraction of the same strain. Sieben et al. [21] conducted biotransformation of agroclavine derivatives with *C. fusiformis* SD-58 to obtain the corresponding derivatives of elymoclavine. The strain was able to transform 1-alkyl-, 1-benzyl-, 1-hydroxymethyl-, 2-halo- and 2,3-dihydroagroclavine and 6-ethyl-6-noragroclavine to the corresponding elymoclavine derivatives. It was shown that the substrate specificity of the agroclavine 17-hydroxylase was high with respect to the 8,9-double bond and to the tertiary state of N-6, but low for variations in the pyrrole partial structure (N-1, C-2, C-3). The N-1 alkylated agroclavines are hydroxylated faster probably due to better penetration of the substances into the cells because of more lipophilic features. Noragroclavine (**3**) and lysergine (**14**) are oxidized by the above system [22] to the corresponding 8α-hydroxyderivatives, i.e. norsetoclavine and setoclavine. This system is highly stereospecific (to 8α oxidation) in contrary to horse radish peroxidase that gives rise to a mixture of 8α and 8β isomers.

For industrial bioconversions of agroclavine to elymoclavine, the high production *C. fusiformis* [23, 24] or selected *C. paspali* [5] strains are the most suitable ones.

The use of immobilized and permeabilized *C. fusiformis* cells for this bioconversion was successfully tested [23].

2.2 Elymoclavine

Most elymoclavine (**2**) bioconversions were focused to its oxidation to lysergic acid (**5**) or paspalic acid (**4**). Similarly as in agroclavine biooxidations, the C-17 oxidation activity is confined to the *Claviceps* genus. Other biosystems, such as fungi [6, 16, 17, 25, 26], bacteria [6, 7, 16] and plant preparations [10, 13, 14, 27, 28] introduce a hydroxy-group at the position C-8 (proxidase reaction) − as in agroclavine − giving rise to penniclavine (**16**) and isopenniclavine (**18**) (Table 2). The production of complicated mixture of penniclavine, isopenniclavine, lysergol [6] and 10-hydroxyderivatives [13, 14, 27] was frequently observed.

Table 2. Elymoclavine bioconversions

Reaction	Product	Conversion system (culture)	Ref.
8-Hydroxylation	Penniclavine	*Aspergillus fumigatus, Mucor corticolus* CBS, *Rhizopus arrhizus* CBS, *Streptomyces rimosus* NRLL-2234, *S. scabies* ATCC-3352, *S.* sp. PCC	[6]
		Penicillium viridicatum	[25]
		Ipomoea (leaves)	[28]
8-Hydroxylation	Isopenniclavine	*Absidia spinosa* ATCC-6648, *Fusarium graminearum* PCC-140	[6]
8-Hydroxylation	Isopenniclavine + penniclavine	*Colletatrislium graminicolum, Cunninghamella echinulata* CBS, *Gibberella zeae* Ull, *Helminthosporium victorie, Rhizopus circinans* CBS, *R. nigricans* PCC-199	[6]
		Streptomyces lipanii, Fusarium niveum, Corticium sasaki	[7, 16]
		Tomatoes, potato sprouts homogenates morning glory seedlings	[10]
		Horse radish peroxidase	[13, 14, 27]
10-Hydroxylation	10-Hydroxy-elymoclavine	Horse radish peroxidase	[13, 14, 27]

Table 2. (continued)

Reaction	Product	Conversion system (culture)	Ref.
8,9-Epoxidation	10-Hydroxy-8,9-epoxyelymoclavine	Horse radish peroxidase	[13, 14, 27]
17-Oxidation	Paspalic acid	*Claviceps* spp. *Claviceps* sp. PCCE1 (*purpurea*?) mycelial fraction	[26] [31]
17-Oxidation	Lysergic acid	*C. purpurea* PEPTY 695/S microsomal fraction *C. paspali* SO 70/5/2	[30] [33]
17-OH-Reduction	Agroclavine	*Aspergillus fumigatus*, *Corticium sasakii*	16, 17, 26]
Reductive opening of D-ring (not proved)	Chanoclavine	*Aspergillus fumigatus* *C. paspali* MG-6	[7, 16] [38]
8,9-Double bond isomerisation	Lysergol	*Beauveria bassiana* PCC-122, *Fusarium lini* ATCC-9593, *F. lycopersici* PCC-141, *F. roseum* PCC-142, *Mucor adventitius* CBS, *Monascus pitosus* IMUR-165, *Rhizopus chinensis* CBS, *Streptomyces parvus* NRRL-B-1456, *S. griseus* (3 strains)	[6]
Isomerisation	Lysergol + isolysergol	*Claviceps* spp.	[18]
Isomerisation + 8-hydroxylation	Lysergole + penniclavine + isopenniclavine	*Aspergillus chevalieri* NRRL-78, *A. fumigatus* (3 strains), *A. niger* ATCC-6277, *A. quadrilineatus* PCC-115, *Botryosporium* sp. PCC-284, *Ceratocystis ulmi*, *Fusarium graminearum* PCC-144, *Mucor corticolus* CBS, *M. flavus* PCC-256, *M. globosus* CBS, *Melanospora destruens* LSHTM BB-168, *Sclerotinia sclerotiorum*, *Nocardia convoluta* PCC-109, *Streptomyces rimosus* NRRL-2234, *S. rubrireticuli* NRRL-B-1484, *S. scabies* ATCC-3352	[6]
17-Oxidation	Ergine	*C. paspali* LI 189 +	[32]
17-Oxidation + derivatization	Lysergic acid α-hydroxyethylamide (mixture of isomeres) *C. paspali* 8063		[34]

Table 2. (continued)

Reaction	Product	Conversion system (culture)	Ref.
17-Oxidation + derivatization	Ergotamine	*C. purpurea* (sclerotia)	[32, 59]
		C. purpurea	[60]
Glycosylation of 17-OH	Elymoclavine-O-β-D-fructofuranoside + elymoclavine-O-β-D-fructofuranosyl-(2→1)-O-β-D-fructofuranoside		
		C. purpurea EK	[35]
Elymoclavine derivatives 1-Alkyl- 1-Benzyl- 1-Hydroxymethyl- 2-Bromo- 6-Ethyl-6-nor-			
17-Oxidation + derivatization	Corresponding lysergic acid α-hydroxyethylamides		
		C. paspali	[33]

Isomerisation of elymoclavine to lysergol (**11**) should be further investigated. Lysergol is a valuable pharmaceutical substrate that is usually produced by alkaline catalyzed isomerisation of elymoclavine in lower yields [29].

Biological reduction of elymoclavine leading to agroclavine is performed by several fungi [7, 16, 17, 26].

Oxidation of elymoclavine to lysergic acid and to its derivatives is practicable only by selected strains of *C. purpurea* and *C. paspali* [26, 30–34]. The enzyme systems responsible for this reaction are localized in the microsomal fraction [30, 31] and belong obviously to the cytochrome P-450 family. This bioconversion was accomplished also on an industrial scale [34]. Derivatives of elymoclavine were analogously converted by *C. paspali* to the corresponding lysergic acid derivatives [33].

Recently, fructosylation of elymoclavine by *C. purpurea* was described [35]. This reaction is mediated by the transfructosylating activity of β-fructofuranosidase in the sucrose-containing media. Beside mono-(**21**) and difructoside (**22**) (Fig. 4), higher fructosides (tri- and tetra-) are probably formed. The reaction is strongly dependent on pH (optimum 6.5) and substrate concentration (opt. 75 g l^{-1} sucrose).

The glycosylating strains produce most of elymoclavine in the glycosidic form that complicates the isolation of elymoclavine. Hydrolysis of fructosides by HCl is not suitable for a large-scale process due to the aggression of the acid solution and losses of elymoclavine. A more elegant method is a bioconversion employing high invertase activity of *Saccharomyces cerevisiae*. At the

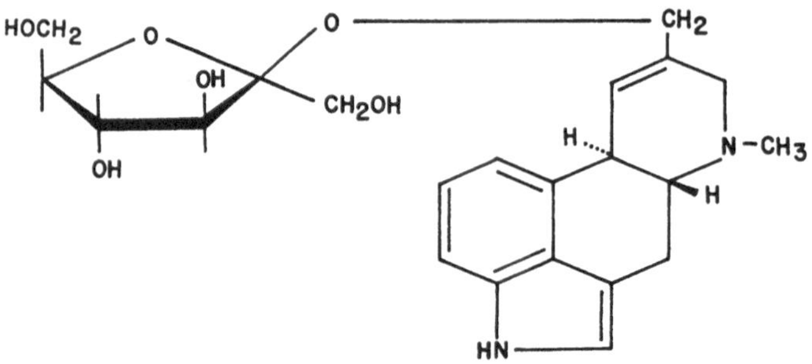

21

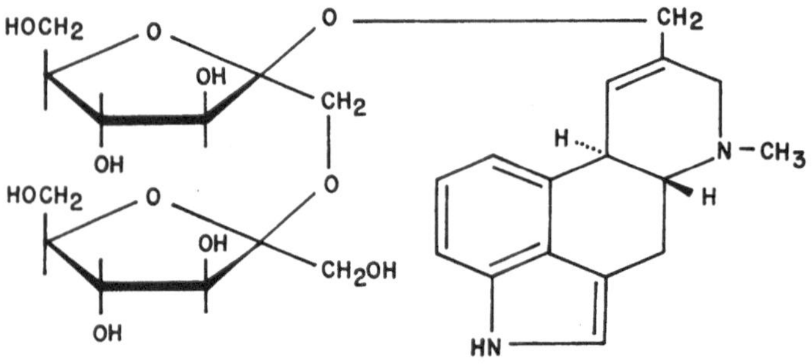

22

Fig. 4. Elymoclavine fructosides
(*21*) elymoclavine-0-β-D-fructoside
(*22*) elymoclavine-0-β-D-fructofuranosyl(2→1)-0-β-D-fructofuranoside

end of the production cultivation, a suspension of baker's yeast is added to the medium (without the product isolation). The hydrolysis is completed within 1 h (37 °C) [35].

2.3 Lysergoles

Lysergoles, i.e. lysergol, lysergine and their isomers occur as minor alkaloids in *Claviceps* cultures. Their biogenetic relations to other alkaloids remain still unclear. Most of lysergoles bioconversion were aimed at discovering their biogenetic relations to other alkaloids.

Lysergol (**11**) and lysergine (**14**) are oxidized by *C. fusiformis* by a similar mechanism as elymoclavine to penniclavine and setoclavine respectively [22]. Still, lysergol is not converted by elymoclavine 17-oxygenase from *C. purpurea* to paspalic acid [30].

Lysergen (13) is converted by various strains of *C. purpurea* to lysergol, isolysergol (12), penniclavine and isopenniclavine [13, 18].

A selected strain *C. purpurea* with high fructofuranosidase activity is able to fructosylate lysergoles by the same mechanism as in the case of elymoclavine [35]. Mono-, di- and tri-glycosides of lysergoles were found by HPLC analysis.

The activity of the glycosylating system towards various lysergoles and elymoclavine was compared to elucidate its substrate specificity [36] (Table 3). Both iso-lysergoles gave very low yield of glycosides. It is probably caused by stabilization of their hydroxymethyl groups by an H-bond (Fig. 5) with the N-6

Table 3. Bioconversions of lysergoles

Reaction	Product	Conversion system (culture)	Ref.
Lysergole			
8-Hydroxylation	Penniclavine	*Claviceps fusiformis* SD-58	[22]
Reduction + isomerization (not proved) Lysergole Dihydrolysergole iso-Lysergole iso-Dihydrolysergole	Agroclavine	*Claviceps* sp. XM5, *Aspergillus fumigatus*, *Penicillium roqueforti*	[16]
17-OH Fructosylation	Corresponding-O-β-D-fructofuranosides		
		Claviceps purpurea EK	[35, 36]
Lysergine			
8-Hydroxylation	Setoclavine	*Claviceps fusiformis* SD-58	[22]
Lysergene			
Oxidative opening of D-ring (not proved)	Chanoclavine	*Penicillium roqueforti*	[16]
8,17-Double bond reduction + 9,10-Double bond isomerization (not proved)	Agroclavine	*Claviceps* sp. XM5, *Aspergillus fumigatus*	[16]
17-Hydroxylation + 8-Hydroxylation	Lysergol, isolysergol penniclavine, isopenniclavine		
		Claviceps purpurea	[16, 18]

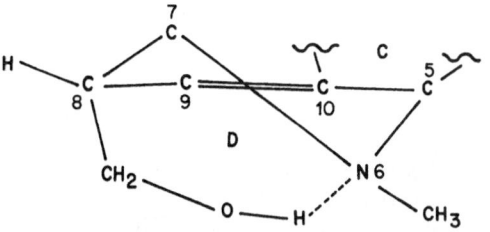

Fig. 5. Stereochemical structure of D-ring of iso-lysergol — see formation of H-bridge with N-6. In lysergole the $-CH_2OH$ group remains free

atom of the D-ring. Considerably high glycosylation rate of elymoclavine (2) can be ascribed to the activation of the hydroxymethyl group by adjacent double bond (allylic) in positions 8, 9.

This bioconversion was performed in a culture of *C. purpurea* with alkaloid production selectively blocked by 5-fluorotryptophan (Table 4). Bioconversion of ergot alkaloids by *Claviceps* strains are often complicated by the production of alkaloids *de novo* that might compete with the added "xeno" alkaloids and make the mixture after conversion rather complex. Strains of *C. purpurea* with the glycosylation activity usually produce high amount of elymoclavine that is glycosylated at a high rate and thus competes with the added lysergoles. 5-Fluorotryptophan blocks the production at the first reaction and the following steps remain active [37]. The growth and differentiation of the culture is not impaired, so the bioconversion proceeds in "physiologically" normal culture and without any endogenous alkaloid production.

Table 4. Lysergoles glycosylation by *C. purpurea* EK

Alkaloid	% of Glycosylated alkaloids	
	Monofructoside	Total glycosides
lysergol	14.7	31.5
dihydrolysergol	16.7	29.2
iso-lysergol	5.7	10.4
iso-dihydrolysergol	5.1	6.9
elymoclavine	36.4	63.6

2.4 Chanoclavine

Chanoclavine-I (23) (Fig. 6) is a common precursor of most ergot alkaloids in the *Claviceps* genus and its conversion to agroclavine and elymoclavine by enzymatic systems of various *Claviceps* strains has been clearly proved [38–40]. Ogunlana

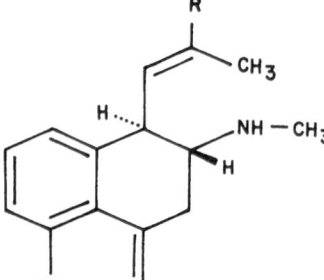

Fig. 6. Secoclavines
(23) chanoclavine-I R = CH_2OH
(24) chanoclavine-I aldehyde R = CHO

et al. [41] reported cyclization of chanoclavine-I to elymoclavine as a sole product by pigeon-liver acetone-powder (+ ATP + MG^{2+}). This is probably the only report referring to this reaction in a system of non-*Claviceps* origin.

The conversion of exogenous chanoclavine by intact mycelia of *C. fusiformis* strain W1 [42] gave chanoclavine-I aldehyde (**24**), elymoclavine and agroclavine. However, compared with agroclavine, the conversion proceed slowly due to the presumably low transport rate of chanoclavine into cells. More polar alkaloids like chanoclavine and elymoclavine enter the cells at a lower rate than less polar agroclavine. Beside these products, also mono- and difructoside of chanoclavine were identified [42]. These glycosides are formed by the same mechanism like the analogous fructosides of elymoclavine and lysergoles.

8,9-Dihydrochanoclavine and isochanoclavine is cycled by *Claviceps* sp. SD-58 to festuclavine (**19**) and pyroclavine (**20**) respectively [61].

2.5 Semisynthetic Clavine Alkaloids

Recent interest in lergotrile (**25**) (Fig. 7) stems from its putative dopaminergic activity and inhibition of prolactin secretion. Toxicity in clinical trials prompted the exploration of microbial transformation for producing less toxic derivatives.

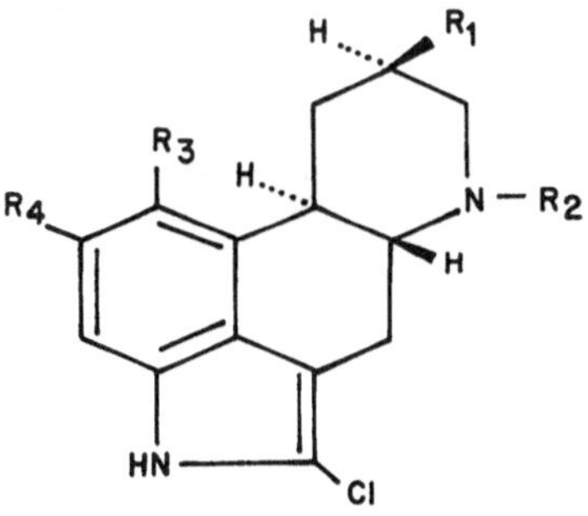

Fig. 7. Semisynthetic ergot drugs and their metabolites
(*25*) Lergotrile $R_1 = CH_2CN$, $R_2 = CH_3$, $R_3 = R_4 = H$
(*26*) nor-lergotrile $R_1 = CH_2CN$, $R_2 = R_3 = R_4 = H$
(*27*) 12-hydroxylergotrile $R_1 = CH_2CN$, $R_2 = CH_3$, $R_3 = OH$, $R_4 = H$
(*28*) 13-hydroxylergotrile $R_1 = CH_2CN$, $R_2 = CH_3$, $R_3 = H$, $R_4 = OH$
(*37*) Pergolide $R_1 = CH_2SCH_3$, $R_2 = CH_2CH_2CH_3$, $R_3 = R_4 = H$

Davis et al. [43] screened nearly 40 organisms for their ability to produce metabolites of lergotrile. Five microorganisms (*Cunninghamela echinulata* UI 3655, *Streptomyces rimosus* ATCC 23 955, *S. platensis* NRRL 2364, *S. spectabilis* UI-C632, *S. flocculus* ATCC 25435) biotransformed lergotrile (**25**) to norlergotril (**26**) by N-demethylation. *S. platensis* exhibited complete conversion, and a preparative-scale incubation was accomplished with a yield of 50%. Additional organisms have been screened for their ability to produce hydroxylated metabolites of lergotrile (**27**, **28**) that has been found in humans and in mammals but these efforts have been unsuccessful [44]. Microbial N-demethylation is important because chemical demethylation of the compounds like lergotrile is somehow difficult [44].

In guinea-pig liver, lergotrile is demethylated and hydroxylated at C-13 and at the nitrile group [45].

3 Bioconversions of Lysergic Acid Derivatives

3.1 Hydrolysis of Lysergic Acid Amides

This group includes simple lysergic acid amides and peptide ergot alkaloids. Most of these substances are used directly in pharmaceutical preparations. However, reports referring to their biotransformation are scarce.

One of the most desired transformations is a hydrolysis of the above substances resulting in free lysergic acid, the substrate for many semisynthetic preparations. Chemical hydrolysis of its derivatives gives low yields (30–40%). Microbial production of lysergic acid is somehow complicated. Enzymatic hydrolysis of peptide alkaloids is still now impracticable. Common proteolytic enzymes (papain, subtilisin, chymotrypsin, thermolysin) do not attack the peptidic bond in ergocryptine (**32**), probably due to steric reasons [46].

Amici et al. [47] reported the hydrolysis of ergine (**7**) and erginine (**8**) by a strain of *C. purpurea* yielding 80–90% lysergic acid. The bioreaction is aerobic. The cultivation medium containing succinate, mannitol and mineral salts at pH 5.2 is supplemented with ergine dissolved in an organic solvent up to the final concentration of 1000 mg l^{-1}.

Ergotoxines producing strains of *C. purpurea* are able to split lysergylaminoacid-methylesters into the corresponding lysergyl-aminoacids as well as into lysergic

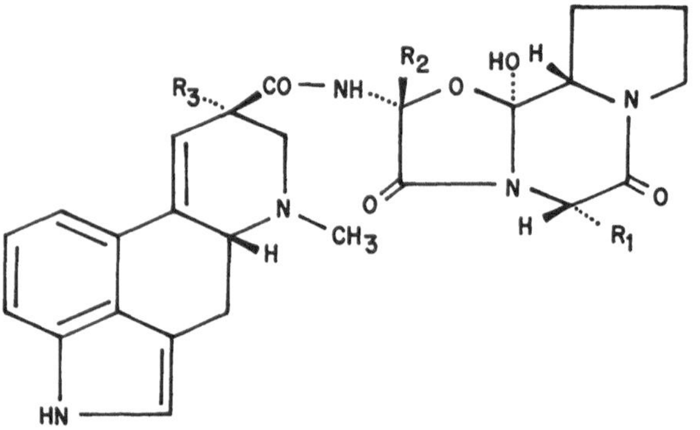

Fig. 8. Peptide ergot alkaloids
(*30*) ergotamine $R_1 = CH_2 - C_6H_5$, $R_2 = CH_3$, $R_3 = H$
(*31*) 8-hydroxyergotamine $R_1 = CH_2 - C_6H_5$, $R_2 = CH_3$, $R_3 = OH$
(*32*) α-ergocryptine $R_1 = CH_2CH(CH_3)_2$, $R_2 = CH(CH_3)_2$, $R_3 = H$

acid and alanine and valine [48]. This reaction proceeds inside the cells, so the transport is involved. Ergotoxines producing strains are more active than the nonproducing strains. Also lysergyl-oligopeptides (d-lysergyl-L-valyl-L-leucine-OMe, d-lysergyl-L-valyl-L-valine-OMe and d-lysergyl-L-valyl-L-proline-OMe) are split into their components after feeding to the mycelium of *C. purpurea* Pepty 695 (ergotoxines producing) [49].

3.2 Oxidation of Lysergic Acid Derivatives

Ergotamine (**30**) (Fig. 8) and ergometrine (**9**) are partially isomerised at C-8 (but not hydroxylated) to ergotaminine and ergometrinine, respectively, by *Psilocybe semperviva* [8] and partially decomposed.

Lysergic acid derivatives can be oxidized, analogously to most clavine alkaloids, at C-8 due to the action of peroxidases. 8-hydroxyergotamine (**31**) has been found in an ergotamine producing *C. purpurea* strain [50]. Oxidation of exogenous ergine (**7**) to 8-hydroxyergine (**29**) by *C. paspali* MG-6 was accomplished by Linhartová et al. [51].

3.3 Biotransformations of LSD and Its Homologs

Lysergic acid amides, most notably lysergic acid diethylamide (LSD) (**33**), are of interest because of their hallucinogenic activity. Microbial metabolites of LSD might serve as more active medical preparations. Other biotransformation products correspond to LSD metabolites found in humans. The two facts led Ishii et al. [52–55] to an extensive study of microbial bioconversions of the LSD and its derivatives.

Initial studies with LSD showed that many cultures were capable of attacking the N-6 and the amide N-alkyl substituents [52–54]. *Streptomyces lavendulae* IFM 1031 demethylated only the N-6 position yielding nor-LSD. Conversely, *Streptomyces roseochromogenes* IFM 1081 attacked only the N-amide alkyl group to yield lysergic acid ethylamide (**34**), lysergic acid ethylvinylamide (**35**), and lysergic acid ethyl 2-hydroxyethylamide (**36**) (Fig. 9). Other *Streptomyces* and *Cunningha-*

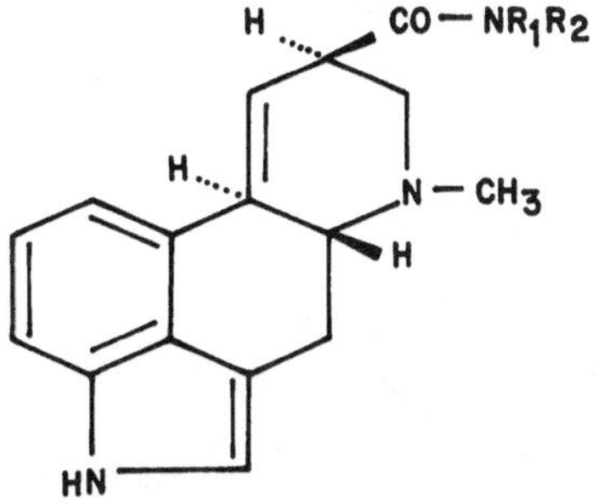

Fig. 9. LSD and its metabolites
(*33*) lysergic acid diethylamide (LSD)
    $R_1 = R_2 = CH_2CH_3$
(*34*) lysergic acid ethylamide $R_1 = H$,
    $R_2 = CH_2CH_3$
(*35*) lysergic acid vinylamide $R_1 = H$,
    $R_2 = CH = CH_2$
(*36*) lysergic acid ethyl 2-hydroxyethylamide
    $R_1 = CH_2CH_3$, $R_2 = CH_2CH_2OH$

mella strains produce all four metabolites. The high degree of substrate stereo-selectivity in *Streptomyces roseochromogenes* was proved by the fact that this organism could not metabolize iso-LSD while, in contrary, *S. lavendulae* yielded iso-nor-LSD.

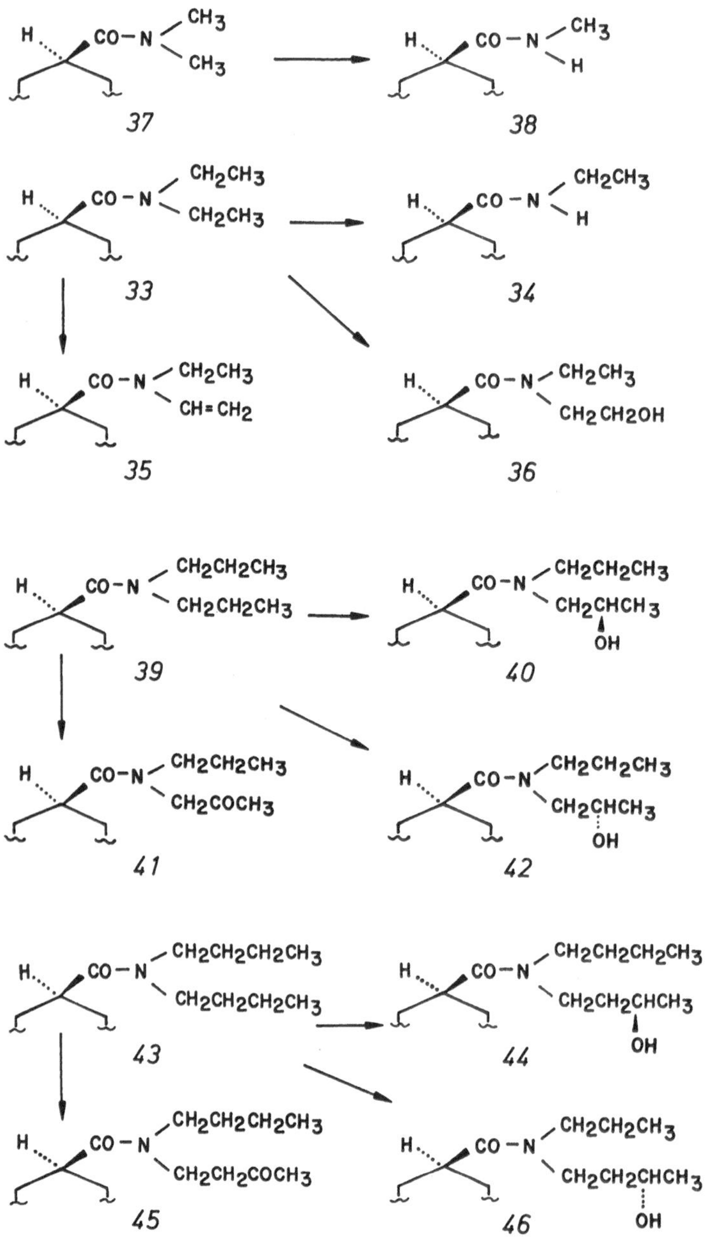

Fig. 10. Metabolism of lysergic acid dialkyl homologs by *Streptomyces roseochromogenes*

C-17-amide dealkylations were examined in detail using a series of lower and higher alkyl homologs of LSD [55] (Fig. 10). The following results were observed: Lysergic acid dimethylamide (37) was only dealkylated to its monomethylamide (38), lysergic acid diethylamide (33) was also dealkylated to yield a monoethylamide (34) and other metabolites like ethylvinylamide (35) and ethylhydroxyethylamide derivatives (36). Neither lysergic acid di-*n*-propylamide (39) nor lysergic acid di-*n*-butylamide (43) were dealkylated, but rather yielded the two epimeric alcohols resulting from ω-1 hydroxylation, i.e. (40) and (42) from (39), and (44) and (46) from (43), as well as the further oxidation products, ketones (41) and (45), respectively. Based on these results, the authors proposed that the chain length regulates the site of oxygenation, with ω-1 hydroxylation occurring if possible. The dimethylderivative (37) simply does not have an ω-1 position so Cα-(methyl)hydroxylation yields the carbinolamide with resulting N-demethylation.

These studies resulted in a proposed active site for the hydroxylase (N-dealkylase) of *S. roseochromogenes* that accounts for the general mode of metabolism of the homologs (Fig. 11). The authors also use the diagram to explain the stereochemical control of the system, based on the observation that one epimeric alcohol predominates in the hydroxylation of (39) or (43). This argument is based on a favored binding of one alkyl group over the other, implying, as the authors explain, non-equivalence of two alkyl groups [55, 56].

Such an "active-site map" should facilitate predicting acceptable substrates and stereochemical outcome on related compounds as accomplished with other classes of substrates in microbial transformations [56].

Microsomes from mammalian liver (+ NADPH + O_2) oxidize LSD to 2-hydroxy-LSD [57].

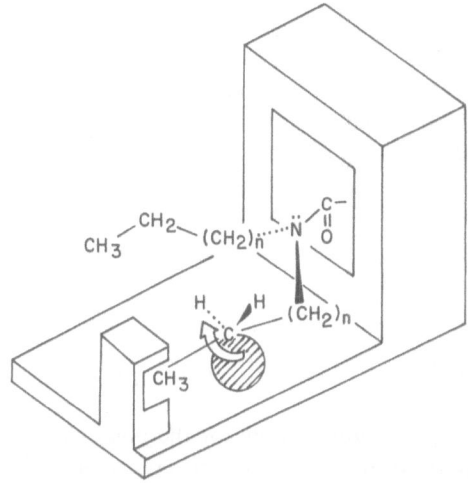

Fig. 11. Proposed active site for the ω-1-hydroxylase of *Streptomyces roseochromogenes* (From Ishii et al. [55])

4 Bioconversion of Ergot Alkaloids as a Tool for Study of Their Metabolism in Mammals

The elucidation of mammalian metabolic pathways is important in attempting to rationalize detoxification, and to evaluate potentially active metabolites. In the case of some ergot alkaloids with strong hallucinogenic activity (e.g. LSD), it is desirable to detect its metabolities for use in antidoping screening and criminalistics. However, such studies have been hampered by a lack of availability of the minor metabolites.

For this reason, Ishii and coworkers [52–55] examined a series of microorganisms and animals for alternative routes of metabolism of LSD (lysergic acid diethylamide) and related compounds. They assumed that the metabolism of these xenobiotics would proceed in an essentially similar manner in both mammals and microorganisms [54]. Both nor-LSD and lysergic acid ethylamide produced by the microbial conversion of LSD are known metabolites of LSD in mammals, and the authors were able to use lysergic acid ethylvinyl amide generated in the microbial studies to determine its presence in mammals.

This work has allowed a high degree of predictability regarding the N-(amide)-alkyl oxidation of lysergic acid amide derivatives using *Streptomyces roseochromogenes* (see above). Analysis of a series of homologs from lysergic acid dimethylamide upto lysergic acid dibutylamide resulted in a proposal for the enzyme active site, as well as rationalization of the resulting metabolites. This study also enabled the preparation of the minor metabolites to be made for further investigation.

Lergotrile (see above) is hydroxylated at various positions in mammalian systems (C-13, C-12, hydrolysis of nitrile group) and demethylated at N-6 [44, 57]. To find a parallel for this reaction and to prepare some standards of the metabolites more than 30 microorganisms have been screened for their hydroxylation ability [43]. Only N-6 demethylation activity has been found in *Streptomyces platensis*.

Microbial systems parallel mammalian metabolism with an other dopaminergic ergoline viz. pergolide (**37**). Metabolism in the mammal centers on the methyl sulfide moiety, which is sequentially oxidized to the sulfoxide and the sulfone. Similarly, *Aspergillus alliaceus* UI-315 catalyzed the same sequential oxidative transformation. In contrast, *Helminthosporium* sp. NRRL 4761 stops at the sulfoxide stage and also catalyzes reduction of the sulfoxide back to the pergolide [56]. No stereoselectivity in sulfoxide formation was observed, in contrast to the high degree of product stereoselectivity often observed in this microbial-type reaction [56, 58].

5 Use of Immobilized Cells in Ergot Alkaloid Bioconversion

Immobilized cells are employed for bioconversions mainly for the possibility of repeated use. There are only two reports dealing with bioconversion of ergot alkaloids by means of immobilized cells.

Biooxidation of agroclavine to elymoclavine is practicable only with some selected *Claviceps* strains. Obtaining the biomass for the bioconversion is hampered by the considerably slow growth of the organism. For this reason, *C. fusiformis* cells immobilized in Ca-alginate were tested [23]. For the preparation of the biocatalyst a mycelium from the inoculation stage was used. The activity of the biocatalyst lasted about 10 cycles (50 days). The biocatalyst can be regenerated in a medium containing biofactors. The efficiency of the conversion can be increased by permeabilization of the immobilized cells by dimethylsulfoxide. The yield of bioconversion in approximately 5-d-cycles was 50–70%, after supplementing with dimethylsulfoxide (1%) the yield was raised to 65–90%.

Claviceps purpurea cells immobilized in Ca-alginate were used for the preparation of ergot alkaloid fructosides [35]. The reason of using immobilized cells is the same as above. In addition, this bioconversion proceeds a longer period (22 days) and free cells tend to autolyse during the process. Here, the mycelium from the production phase with high glycosylation activity was used. The conversion rate drops from 62% in the first cycle to 24% in the third cycle. The stability of this immobilized system is rather lower.

While the biooxidation of agroclavine proceeds intracellularly, the enzymes responsible for glycosylation are localized in periplasmatic space. Thus there are no transport limitations in this case.

6 Prospects for the Future

Production of most desired ergot substances, e.g. elymoclavine, paspalic acid and lysergol, used mainly for semisynthetic ergot preparations, is rather complicated. Bioconversion of their precursors may improve the effectivity of their production.

The most important ergot alkaloid bioconversion from the industrial point of view is the oxidation of the ergoline molecule at C-17. The precursors in the lower oxidation degree, i.e., agroclavine and chanoclavine, can be produced in considerably high amounts. Elymoclavine acts as a feedback inhibitor of its production in *Claviceps*, so its yield in batch cultivation is limited. Another highly desired reaction is the bioconversion of clavine alkaloids to paspalic acid. Both these biooxidations seem to be performed only by the organisms within the *Claviceps* genus. Although some work on these problems has already been done, both processes should be amended and adopted for large scale applications.

Isomerisation of elymoclavine to lysergol is up to now performed chemically with rather low yields. Although there are some preliminary reports on bioconversion of elymoclavine to lysergol [16], an applicable method is still lacking.

In some cases, lysergic acid is produced by alkalic hydrolysis of the peptide ergot alkaloids or lysergic acid derivatives. Harsh conditions of this chemical reaction cause a drop in the yields due to the decomposition of the ergoline skeleton. This could be another challenge for a further search for bioconversion methods.

Recently, immunomodulatory activity of clavine alkaloids (elymoclavine, dihydrolysergol etc.) has been discovered [62, 63]. This activity is higher in the case of the fructosylated alkaloids. However, it could be expected that the glucosylated or mannosylated alkaloids would be even more effective. Thus these glycosylations performed by some biosystem might give rise to more effective drugs.

Immobilized cells are applicable for most bioconversions. Mixed cultures, either free or immobilized, could simplify performance of these bioreactions. However, meeting different nutritional requirements of mixed cultures is a problem that must be overcome.

7 Acknowledgements

I wish to thank Professor D. Gröger from the Inst. of Plant Biochemistry (Halle, Saale, Germany) for critical reading of the manuscript. I am also indebted to Dr. S. Pažoutová, Institute of Microbiology, Prague for many discussions and generous cooperation and to Drs. Z. Malinka and P. Harazim, Galena Pharmaceutical Works (Opava, Czechoslovakia), for providing information on industrial applications of ergoline bioconversions.

Abbreviations of Culture Collections

ATCC   = American Type Culture Collection, Washington, D.C., U.S.A.
CBS   = Centralbureau voor Schimmelculturen, Holland
IMUR   = Instituto de Micologia Universitad Recife, Recife, Brasil
LSHTM = London School of Tropical Medicine, Great Britain
NRRL   = Northern Regional Research Laboratory, Peoria, Illinois, U.S.A.
PCC   = Purdue Culture Collection, Lafayette, Indiana, U.S.A.
Ull   = Collection of Dr. A. J. Ullstrup, Purdue University, Lafayette, Indiana, U.S.A.

References

1. Fanchamps A (1979) J Pharmacol 10: 567
2. Gröger D (1972) Ergot alkaloids. In: Ajl SJ, Kadis S, Montie TC (eds) Microbial toxins, vol. 8, Academic, New York, p 321
3. Floss HG (1976) Tetrahedron 32: 873
4. Kobel H, Sanglier J-J (1986) Ergot alkaloids. In: Rehm H-J, Reed G, (eds) Biotechnology, vol 4, VCH, Weinheim, p 569
5. Flieger M, Linhartová R, Řeháček Z, Sajdl P, Stuchlík J, Malinka Z, Harazim P, Cvak L, Břemek J (1989) Czechoslovak Pat Appl PV 6958–89
6. Béliveau J, Ramstad E (1966) Lloydia 29: 234
7. Yamatodani S, Kozu Y, Yamada S, Abe M (1962) Ann. Repts. Takeda Res Lab 21: 88, viz Chem Abstr (1963) 59: 3099c
8. Brack A, Brunner R, Kobel H (1962) Helv Chim Acta 45: 278
9. Davis PJ (1982) Microbial transformation of alkaloids. In: Rosazza JP (ed) Microbial transformation of bioactive compounds, vol 11, CRC Press, Inc, Boca Raton, Florida, p 67

10. Taylor EH, Goldner KJ, Pong SF, Shough HR (1966) Lloydia 29: 239
11. Taylor EH, Shough HR (1967) Lloydia 30: 197
12. Shough HR, Taylor EH (1969) Lloydia 32: 315
13. Chan-Lin WN, Ramstad E (1967) Lloydia 30: 284 P
14. Shough HR, Pong SF, Taylor EH (1967) Lloydia 30: 284 P
15. Wilson EJ, Ramstad E, Janson I, Orrenius S (1971) Biochim Biophys Acta 252: 348
16. Abe M (1966) Abhandl Deut Akad Wiss Berlin, Kl Chem No 3, p 393
17. Abe M, Yamatodani S, Yamano T, Kozu Y, Yamada S (1963) Agric Biol Chem 27: 659
18. Agurell S, Ramstad E (1962) Arch Biochem Biophys 98: 457
19. Hsu JC, Anderson JA (1971) Biochim Biophys Acta 1971, 230, 518
20. Kim I-S, Kim S-U, Anderson JA (1981) Phytochemistry 20: 2311
21. Sieben R, Philippi U, Eich E (1984) J Nat Prod 47: 433
22. Eich E, Sieben R (1985) Planta Medica 1985: 282
23. Křen V, Břemek J, Flieger M, Kozová J, Malinka Z, Řeháček Z (1989) Enzyme Microb Technol 11: 685
24. Břemek J, Křen V, Řeháček Z, Sajdl P, Spáčil J, Malinka Z, Kozová J, Krajíček A, Flieger M, Pilát P (1989) Czechoslovak Pat AO 257311
25. Tyler VE, Erge D, Gröger D (1965) Planta Medica 13: 315
26. Sebek OK (1983) Mycologia 75: 383
27. Chang-Lin WN, Ramstad E, Taylor EH (1967) Lloydia 30: 202
28. Gröger D (1963) Planta Medica 11: 444
29. Schreier E (1958) Helv Chim Acta 41: 1984
30. Maier W, Schumann B, Gröger D (1988) J Basic Microbiol 28: 83
31. Kim S-U, Cho Y-J, Floss HG, Anderson JA (1983) Planta Medica 48: 145
32. Mothes K, Winkler K, Gröger D, Floss HG, Mothes K, Weygand F (1962) Tetrahedron Lett 1962: 933
33. Philippi U, Eich E (1984) Planta Medica 50: 456
34. Harazim P, Malinka Z, Stuchlík J, Cvak L, Břemek J, Flieger M, Linhartová R, Řeháček Z, Sajdl P (1989) Czechoslovak Pat Appl PV 6959-89
35. Křen V, Flieger M, Sajdl P (1990) Appl Microbiol Biotechnol 32: 645
36. Křen V, Flieger M, Svatoš A, Sajdl P (1989) Glycosylation of ergot alkaloids. In: Černý M, Drašar P (eds) Eurocarb V, 5th European Symp on Carbohydrates, 21-25 Aug 1989. Prague, p C-32
37. Pažoutová S, Křen V, Sajdl P (1990) Appl Microbiol Biotechnol 33: 330
38. Sajdl P, Řeháček Z (1975) Folia Microbiol 20: 365
39. Erge D, Maier W, Gröger D (1973) Biochem Physiol Pflanzen 164: 234
40. Ogunlana EO, Wilson BJ, Tyler VE, Ramstad E (1970) Chem Commun 1970: 775
41. Ogunlana EO, Ramstad E, Tyler VE, Wilson BJ (1969) Lloydia 32: 524 P
42. Flieger M, Křen V, Zelenkova NF, Sedmera P, Novák J, Sajdl P (1990) J Nat Prod 53: 171
43. Davis PJ, Glade JC, Clark AM, Smith RV (1979) Appl. Environ Microbiol 38: 891
44. Smith RV, Rosazza JP (1982) Microbial transformation as means of preparing mammalian drug metabolites. In: Rosazza JP (ed) Microbial Transformation of Bioactive Compounds. CRC Press, Inc, Boca Raton, Florida, p 1
45. Parli CJ, Smith B (1975) Fed Proc 34: 813
46. Křen V, unpublished results
47. Amici AM, Minghetti A, Spalla C, Tonolo A (1964) Fr Pat 1362876
48. Maier W, Erge D, Gröger D (1974) Biochem Physiol Pflanzen 165: 479
49. Maier W, Baumert A, Gröger D (1978) Biochem Physiol Pflanzen 172: 15
50. Krajíček A, Trtík B, Spáčil L, Sedmera P, Vokoun J, Řeháček Z (1980) Czech Chem Commun 44: 2255
51. Bumbová-Linhartová R, Flieger M, Sedmera P, Zima J (1991) Appl Microbiol Biotechnol 34: 703
52. Nivaguchi T, Nakahara Y, Inoue T, Hayashi M, Ishii H (1975) In: Proc of 19th Symp Chem Nat Prod 1975, Hiroshima, p 235
53. Ishii H, Hayashi M, Niwaguchi T, Nakahara Y (1979) Chem Pharm Bull 27: 1570

54. Ishii H, Niwaguchi T, Nakahara Y, Hayashi MJ (1980) Chem Soc Perkin Trans I 1980: 902
55. Ishii H, Hayashi M, Niwaguchi T, Nakahara Y (1979) Chem Pharm Bull 27: 3029
56. Davis PJ (1984) Natural and semisynthetic alkaloids. In: Rehm H-J, Reed G (eds) Biotechnology, vol 6a, Kieslich K (ed) Biotransformations, Verlag Chemie, Basel, p 207
57. Axelrod J, Brady RO, Witkop B, Evans EW (1956) Nature 178: 143
58. Auret BJ, Boyd DR, Brun F, Greene RME (1981) J Chem Soc Perkin Trans I 1981: 930
59. Winkler K, Mothes K (1962) Planta Medica 10: 208
60. Maier W, Erge D, Schumann B, Gröger D (1981) Biochim Biophys Res Commun 99: 155
61. Johne S, Gröger D, Lier D, Voigt R (1972) Pharmazie 27: 801
62. Fišerová A, Pospíšil M, Kubrycht J, Huan N, Šterzl J (1989) The peripheral molecules of LAK cells and their role in the spontaneous cytotoxicity. In: 7 Int Congr Immunol, 30 July–5 Aug 1989. W. Berlin, Gustav Fischer Verlag Stuttgart, p 686
63. Fišerová A, Flieger M, Pospíšil M, Sajdl P, Táborský O, Cvak L, Stuchlík J (1990) Czechoslovak Pat Appl PV 10990-90

Author Index Volumes 1—44

Marison, I. W. see von Stockar, U. Vol. 40, p. 93

Markkanen, P. see Enari, T. M. Vol. 5, p. 1

Marsili-Libelli, St.: Modelling, Identification and Control of the Activated Sludge Process. Vol. 38, p. 89

Martin, J.-F.: Control of Antibiotic Synthesis by Phosphate. Vol. 6, p. 105

Martin, J. F., Liras, P.: Enzymes Involved in Penicillin, Cephalosporin and Cephamycin Biosynthesis: Vol. 39, p. 153

Martin, P. see Zajic, J. E. Vol. 22, p. 51

McCracken, L. D., Gong, Ch.-Sh.: D-Xylose Metabolism by Mutant Strain of Candida sp. Vol. 27, p. 33

Miller, O. A., Menozzi, F. D., Dubois, D.: Microbeads and Anchorage-Dependent Eukaryotic Cells: The Beginning of a New Era in Biotechnology. Vol. 39, p. 73

Meijer, E. M. see Kamphuis, J. Vol. 42, p. 133

Menozzi, F. D. see Miller, O. A. Vol. 39, p. 73

Merchuk, J. C.: Shear Effects on Suspended Cells. Vol. 44, p. 65

Messing, R. A.: Carriers for Immobilized Biologically Active Systems. Vol. 10, p. 51

Metz, B., Kossen, N. W. F., van Suijidam, J. C.: The Rheology of Mould Suspensions. Vol. 11, p. 103

Misawa, M.: Production of Useful Plant Metabolites. Vol. 31, p. 59

Miura, Y.: Submerged Aerobic Fermentation. Vol. 4, p. 3

Miura, Y.: Mechanism of Liquid Hydrocarbon Uptake by Microorganisms and Growth Kinetics Vol. 9, p. 31

Moo-Young, M., Blanch, H. W.: Design of Biochemical Reactors Mass Transfer Criteria for Simple and Complex Systems. Vol. 19, p. 1

Moo-Young, M. see Scharer, J. M. Vol. 11, p. 85

Morandi, M., Valeri, A.: Industrial Scale Production of β-Interferon. Vol. 37, p. 57

Müller, R. see Syldatk, Ch. Vol. 41, p. 29

Munack, A. see Luttmann, R. Vol. 32, p. 95

Nagai, S.: Mass and Energy Balances for Microbial Growth Kinetics. Vol. 11, p. 49

Nagamune, T. see Endo, I. Vol. 42, p. 1

Nagatani, M. see Aiba, S. Vol. 1, p. 31

Nakajima, H. see Tanaka, A. Vol. 42, p. 97

Nakamura, I. see Kamihara, T. Vol. 29, p. 35

Neubeck, C. E. see Faith, W. T. Vol. 1, p. 77

Neirinck, L. see Schneider, H. Vol. 27, p. 57

Nyeste, L., Pécs, M., Sevella, B., Holló, J.: Production of L-Tryptophan by Microbial Processes, Vol. 26, p. 175

Nyiri, L. K.: Application of Computers in Biochemical Engineering. Vol. 2, p. 49

Ochsner, U. see Reiser, J. Vol. 43, p. 75

O'Driscoll, K. F.: Gel Entrapped Enzymes. Vol. 4, p. 155

Oels, U. see Schügerl, K. Vol. 7, p. 1

Ohshima, T., Soda, K.: Biochemistry and Biotechnology of Amino Acid Dehydrogenases. Vol. 42, p. 187

Okabe, M. see Aiba S. Vol. 7, p. 111

Olson, N. F. see Finocchiaro, T. Vol. 15, p. 71

Onken, U., Liefke, E.: Effect of Total and Partial Pressure (Oxygen and Carbon Dioxide) on Aerobic Microbial Processes. Vol. 40, p. 137